A COURSE
IN MODERN ALGEBRA

A COURSE IN MODERN ALGEBRA

PETER HILTON
Battelle Seattle Research Center and Case Western Reserve University

YEL-CHIANG WU
Oakland University

A WILEY-INTERSCIENCE PUBLICATION

JOHN WILEY & SONS, New York · London · Sydney · Toronto

Library of Congress Cataloging in Publication Data:

Hilton, Peter John.
  A course in modern algebra.

  (Pure and applied mathematics)
  "A Wiley-Interscience publication."
  Bibliography: p.
  1. Algebra. I. Wu, Yel-Chiang, 1939– joint
author. II. Title.

QA155.H54   512   73-18043
ISBN 0-471-39967-1

Printed in the United States of America

10 9 8 7 6 5 4 3 2 1

PREFACE

This book is intended to familiarize the student with a characteristic development and mode of reasoning in modern algebra. It does not aim to be in any sense comprehensive—especially not in regard to a specific mathematical theory such as group theory—but it should enable the student to feel at home with various algebraic theories. The actual mathematical content is such that, at the end, he will have been brought to the frontier of homological algebra and a little beyond. The only formal prerequisite for embarking on this text is a familiarity with linear algebra. Nevertheless, it is expected that most students will have been introduced to several basic algebraic structures—group, abelian group, ring, field—in an earlier algebra course.

What is new about this course, as an undergraduate course, is the point of view. Rather than concentrate on a single mathematical domain such as group theory, ring theory, or commutative group theory, we first describe a number of different algebraic domains and then emphasize the similarities and differences between them. Naturally, to discuss different mathematical domains simultaneously, we had to adopt the terminology of categories and functors. However, we emphasize that this is not a textbook in the theory of categories. We introduce the language to the extent that it is needed, and we prove certain very elementary results in category theory, again taking the theory only as far as the applications dictate. Of course, since we discuss various algebraic domains, it also follows that this is not a textbook in group theory, or any other specific algebraic theory.

Examples of the type of question to which we direct attention are the categorical concepts of *product* and *short exact sequence*. The concept of a product specializes in set theory to the Cartesian product; in group theory to the direct product; in abelian group theory to the direct sum; and in module theory again to the direct sum. On the other hand, the duality principle leads us to look at *coproducts*, and these take very different forms in the various categories under consideration. For sets we have the disjoint union; in group theory we have the free product; in abelian group theory and in module theory we again have the direct sum. Thus categorical properties of products will be reflected in elementary universal properties of both products and coproducts in the various categories, although the realizations of these concepts in the different categories will appear in quite different forms.

With regard to short exact sequences, we know that, for abelian groups, whenever we are given a group A and a subgroup A', we may always form a quotient group $A'' = A/A'$, and we have a short exact sequence

$$A' \overset{\mu}{\rightarrowtail} A \overset{\epsilon}{\twoheadrightarrow} A'',$$

where μ is the embedding and ϵ is the projection. A similar statement holds in module theory, but the homomorphisms μ, ϵ will, of course, now be module homomorphisms. In group theory, we only get a short exact sequence in this way when the subgroup is normal. A natural question to ask in connection with the short exact sequence given above is whether it splits, either on the left or on the right. That is, we ask whether there exists $\kappa : A \rightarrow A'$ with $\kappa\mu = 1$, and whether there exists $\nu : A'' \rightarrow A$, with $\epsilon\nu = 1$. In the theory of vector spaces, both of κ, ν always exist. In the theory of abelian groups (and of modules) κ exists if and only if ν exists, but it may very well happen that both fail to exist. In the theory of groups, ν exists if κ exists, but not conversely. In all these theories, κ exists if and only if A is equivalent to the product of A' and A''. (All these details are, of course, to be found in this book.) Thus the universal point of view that we are adopting leads us to pose the same question in various categories, but the answers will, as indicated above, vary. We regard it as important that the student should appreciate that certain fundamental concepts reappear in various parts of algebra (and indeed in various parts of mathematics beyond algebra), so that we are very often posing the same form of question. However, only the most trivial questions will admit answers that are independent of the particular mathematical context, that is, of the particular category under study.

We now outline the scope of the book. After a brief recall of a few fundamental notions of set theory, we deal, in Chapter 1, with the theory of groups. We do not require that the reader should have met the notion of a group before, although this would in fact put him in a favorable position. Naturally, we do not take the theory very far, but we are guided by our plan to stress universal concepts. Thus we place particular emphasis in the chapter on homomorphisms, direct and free products, free groups, presentations of groups, and exact sequences.

Chapter 2 is concerned with the theory of abelian groups. Here the same principal items occur as in Chapter 1, but of course the theorems are not the same. Indeed, already in this chapter we make a special point of being explicit about the differences between the two theories. We also introduce, in the last section of the chapter, the very important concept of the tensor product of abelian groups. This concept has no parallel in the theory of groups for a very sufficient categorical reason.

Chapters 1 and 2 in a sense set the stage by establishing the point of view of the entire book. Chapter 3 then develops the appropriate language for making the comparisons and the unifications. It describes the basic universal concepts of *category, functor, natural transformation,* and *duality*. The rest of the chapter is devoted to certain auxiliary categorical notions that play a prominent role in the sequel. Our aim is to introduce the notions and to prove those universal properties of these concepts that are used later or have, implicitly, been already used. We consider it very important to separate these universal or trivial parts of a piece of mathematical reasoning* from the deeper and more delicate parts. Not only does this give the student a better insight into the strategy of mathematical argument; it also avoids the superfluous repetition of the same trivial argument in a number of very different contexts.

In Chapter 4 we take up the theory of modules. We regard the concept of a module as a generalization of that of abelian group. It may also of course be regarded as a generalization of that of a vector space, and consequently the contrast can be made between either of these more elementary theories and the theory of modules. Naturally, certain concepts from abelian group theory take on an additional subtlety when the ground ring R is generalized from being simply the ring of integers, Z. For example, the concepts of free module and projective module coincide for abelian groups but do not coincide in the general case.

In Chapter 5 we once again impose constraints on the ground ring R over which our modules are defined. However, we do not simply return to the ring Z! Instead, we look at two generalizations of this ring, namely, at *principal ideal domains* and at *Noetherian rings*. We also study a further generalization of principal ideal domains—*unique factorization domains*. In this chapter we are interested in the ground rings themselves, as rings, as well as in the theory of modules. Thus we not only discuss modules over principal ideal domains and Noetherian rings, but also the properties of the rings themselves. In particular, in connection with unique factorizations domains, we do not have anything to say in the chapter about any special properties of the theory of modules over such domains.

Chapter 6 is a brief supplementary chapter on semisimple rings, which serves to illustrate the role of abstraction and generalization in mathematics and utilizes many of the results obtained in Chapter 5. The content of this chapter could certainly be omitted on first reading.

* Trivial arguments in mathematics are not to be avoided, but they should be recognized as such!

Chapter 7 crosses the frontier into homological algebra. Here we take advantage of the preparation that has been made, in terms of the module theory developed and the universal language employed, to define the two most fundamental *derived functors* of homological algebra, namely Ext and Tor. We prove the most basic properties of these functors, again benefiting from having the requisite language at our disposal. It may be claimed that the student who has read this chapter will have sufficient *algebraic* background to take a first course in algebraic topology. He will also be thoroughly prepared to embark on a text in homological algebra—for example, MacLane's *Homology* or Hilton-Stammbach's *A Course in Homological Algebra*. Indeed, his preparation will have gone well beyond that which could be said to be strictly necessary for reading either of these two texts.

Chapter 7 is followed by a list of the most important symbols used in the book, a bibliography, and an index. The bibliography lists books to which we have made reference in the text and books that could be read after completing the present study.

This book arose out of a course given by one of us, and the experience in that course suggested that the students responded well to the point of view. They were certainly not sophisticated students, and, for them, the course was emphatically not retrospective. Thus, although the text could be used to give an integrated look at undergraduate algebra, it could serve as an introduction to the various mathematical theories treated in the chapters.

We express our appreciation to Ms. Beatrice Shube for her encouragement and interest during the development of this book, and to Ms. Sandra Smith for her skillful, willing, and always cheerful assistance in the preparation of the manuscript.

PETER HILTON

YEL-CHIANG WU

Seattle, Washington
Cleveland, Ohio
Rochester, Michigan
August 1973

CONTENTS

A COURSE
IN MODERN ALGEBRA

PREREQUISITE SET THEORY

This book assumes that the reader is familiar with such terms as sets, functions, composition of functions, injective and surjective functions, the intersection and union of sets, and the complement of a subset in a given set. In the following paragraphs we collect some additional facts that are needed.

Let $\{A_i\}_{i \in I}$ be a collection of sets A_i; their *Cartesian product* is the set

$$\prod_{i \in I} A_i = \left\{ f : I \to \bigcup_{i \in I} A_i \mid f(i) \in A_i \quad \text{for all} \quad i \in I \right\}.$$

If $f(i) = a_i \in A_i$, we write $\{a_i\}_{i \in I}$, or simply $\{a_i\}$, for f, and call a_i the *i*th *component* of $\{a_i\}$. The natural functions $p_i : \prod_{i \in I} A_i \to A_i, p_i(\{a_i\}) = a_i$, are called *projections*. If $I = (1, 2, \ldots, k)$ is a finite set, we write $\prod_{i \in I} A_i = A_1 \times A_2 \times \cdots \times A_k$.

A *relation* R on a set A is a subset of $A \times A$. We write aRb if $\{a, b\} \in R$. A relation R is *reflexive* if aRa for all $a \in A$, *symmetric* if $(aRb \Rightarrow bRa)$, and *transitive* if $(aRb, bRc \Rightarrow aRc)$. Then R is an *equivalence relation* if it is reflexive, symmetric, and transitive. We often denote an equivalence relation by $\sim$, writing $a \sim b$ for aRb. If $\sim$ is an equivalence relation on A, the *equivalence class* containing a is the set

$$[a] = \{ b \mid b \sim a, b \in A \}.$$

PROPOSITION. *If $\sim$ is an equivalence relation on* A, *then*

$$\text{(i)} \quad [a] \cap [b] \neq \phi \Leftrightarrow [a] = [b]$$

$$\text{(ii)} \quad A = \bigcup_{a \in A} [a].$$

Thus the set of *distinct* equivalence classes, $[a_i], i \in I$, partitions A, and the set $\{a_i\}_{i \in I}$ is called a *complete set of representatives of the equivalence classes*. The set of all equivalence classes is called the *quotient set of A*

1

modulo $\sim$, and is denoted by $A/\sim$. There is a natural *projection* $\pi: A \to A/\sim$ which puts each element of A into its equivalence class.

A set A is *preordered* if there is given a relation $\leqslant$ on A which is reflexive and transitive. It is (*partially*) *ordered* if, in addition, it is *anti-symmetric*, that is, $(a \leqslant b, b \leqslant a \Rightarrow a = b)$. If A is preordered by $\leqslant$ we write $a < b$ to mean $a \leqslant b, a \neq b$; $a \geqslant b$ to mean $b \leqslant a$; and $a > b$ to mean $b < a$. We call $a \in A$ an *upper bound* (a *lower bound*) of $B \subseteq A$ if $b \leqslant a$ ($b \geqslant a$) for every $b \in B$. We call $a \in A$ *maximal* (*minimal*) if $b > a$ ($b < a$) is false for every $b \in A$. The set A is a *chain* or *totally ordered* if it is ordered and if, for all $a, b \in A$, one of the three possibilites $a < b, a = b, a > b$ must occur. An ordered set A is *inductive* if it is nonempty and every subset of A which is a chain has an upper bound. Finally, a chain A is *well-ordered* if every nonempty subset of A has a *first* (least) element. In this book we sometimes need to assume the following equivalent axioms.

ZORN'S LEMMA. *Every inductive partially ordered set has a maximal element.*

WELL-ORDERING PRINCIPLE. *Every set can be well-ordered.*

For a proof of equivalence, see J. L. Kelley, *General Topology*, Van Nostrand, 1955.

1

GROUPS

The notion of a group arose historically out of the attempt to extend the classical procedures for solving polynomial equations of degree $\leqslant 4$ to equations of higher degree. It thus came about that groups were originally conceived as groups of permutations, that is, groups of transformations of (finite) sets. In Klein's *Erlanger Programm* groups again appear as groups of transformations and geometry is defined as the study of invariants (of subsets of real coordinate space, for example) under a given group of transformations.

Today the notion of a group has been abstracted from its concrete realizations, and group theory is a well-established independent mathematical discipline. However, the ubiquity of the group concept in mathematics, extending far beyond the theory of equations and geometry—to the calculus, functional analysis, topology, and theoretical physics—ensures that results of group theory have far-reaching implications.

In this chapter we establish some basic facts about groups, set up devices to compare different groups, and describe some standard procedures for constructing new groups out of given groups.

1. THE DEFINITION OF A GROUP

We say that a pair $(G, *)$, consisting of a set G and a binary operation $*$ on G, that is, a function $* : G \times G \rightarrow G$, is a *group* if the following three axioms are verified[†]:

$\mathbf{G_1}$: (*Associativity*) For all a, b, c in G,

$$(a * b) * c = a * (b * c).$$

[†]Note that we write $a * b$ instead of $*(a, b)$, $a, b \in G$. This notation is suggested by the examples of groups cited below.

3

G_2: (*Existence of neutral element*) There exists e in G such that, for all a in G,

$$a * e = e * a = a.$$

G_3: (*Existence of inverses*) To each a in G, there exists $\bar{a}$ in G, such that

$$a * \bar{a} = \bar{a} * a = e.$$

The group $(G, *)$ is said to be *commutative* if the following axiom is also verified:

G_4: (*Commutativity*) For all a, b in G

$$a * b = b * a.$$

EXAMPLES

1. $G = \mathbf{Z}$, the set of integers; $* = +$, addition.

2. $G = \mathbf{Q}^+$, the set of positive rationals; $* = \times$, multiplication.

3. $G = m\mathbf{Z}$, the set of integers divisible by m; $* = +$, addition.

4. $G = \mathbf{Z}_m$, the set of residue classes modulo m; $* = +$, addition.

5. $G = $ set of rigid motions in the plane; $* = $ composition of functions.

6. $G = P_S$, the set of permutations of a set S; $* = $ composition of functions.

The reader should observe that Examples 1, 2, 3, and 4 are commutative, while Example 5 is not, and Example 6 is commutative if and only if S has $\leqslant 2$ elements.

We now establish some elementary properties of groups.

PROPOSITION 1.1. *Let* $(G, *)$ *be a group. Then the element* e *postulated in* G_2 *is unique; and the element* $\bar{a}$ *postulated in* G_3 *is uniquely determined by* a.

PROOF. Suppose e' is also a neutral element for $(G, *)$. Then $e' * e = e$, since e' is a neutral element, and $e' * e = e'$, since e is a neutral element. Thus $e = e'$.

Now suppose that $\tilde{a}$ is also an inverse of a. Then $\bar{a} * (a * \tilde{a}) = \bar{a} * e$, since $\tilde{a}$ is an inverse of a, whence $\bar{a} * (a * \tilde{a}) = \bar{a}$, by G_2. On the other hand, $(\bar{a} * a) * \tilde{a} = e * \tilde{a}$, since $\bar{a}$ is an inverse of a, whence $(\bar{a} * a) * \tilde{a} = \tilde{a}$, by G_2. Since $(\bar{a} * a) * \tilde{a} = \bar{a} * (a * \tilde{a})$, by G_1, we infer that $\bar{a} = \tilde{a}$.

Actually, one can prove more. Let us replace $\mathbf{G_2}$ by $\mathbf{G_{2r}}$.

$\mathbf{G_{2r}}$: There exists e such that, for all a in G,

$$a * e = a.$$

Let us replace $\mathbf{G_3}$ by $\mathbf{G_{3r}}$.

$\mathbf{G_{3r}}$: To each a in G, there exists $\bar{a}$ in G, such that

$$a * \bar{a} = e.$$

We leave to the reader the proof of

PROPOSITION 1.2. $\mathbf{G_1, G_{2r}, G_{3r}}$, *together imply* $\mathbf{G_1, G_2, G_3}$. [Hint: Compute $(\bar{a} * (a * \bar{a})) * \bar{\bar{a}}$ and $\bar{a} * (a * (\bar{a} * \bar{\bar{a}}))$.]

This proposition is significant, since we do not insist that a group be commutative. It is of practical importance, since it simplifies the test as to whether a given pair $(G, *)$ is a group.

PROPOSITION 1.3. *Let* $(G, *)$ *be a group. Then*

$$(i) \qquad \bar{\bar{a}} = a;$$

$$(ii) \qquad \overline{a * b} = \bar{b} * \bar{a}.$$

PROOF.

(i) $\bar{\bar{a}}$ is, by Proposition 1.1, that element b satisfying

$$\bar{a} * b = b * \bar{a} = e.$$

But the element a satisfies these equations.

(ii) We must check that $(a * b) * (\bar{b} * \bar{a}) = e$, $\bar{b} * \bar{a} * (a * b) = e$ (actually, by Proposition 1.2, it is enough to check the former, and this is all we shall do).

Now

$$(a * b) * (\bar{b} * \bar{a}) = ((a * b) * \bar{b}) * \bar{a} = (a * (b * \bar{b})) * \bar{a} = (a * e) * \bar{a} = a * \bar{a} = e.$$

We next introduce some convenient notations. First, if the binary operation $*$ is given by the context, we permit ourselves to write G instead of $(G, *)$. Thus we talk of "the group G." However, this convention must be used with care, since a group is *not* determined by its underlying set; the symbol $*$ may only be suppressed when the context makes perfectly clear what the group operation is. (Note where $*$ is retained in Section 2.)

Our next notational convention is very different in nature. Very many examples of groups, including the most familiar ones, are drawn from our number system (see Examples 1 to 4). They thus involve the operations of addition or multiplication. Now, in Example 2 the role of neutral element is played by the number 1. and the inverse of the rational number a is just the rational number a^{-1}. Moreover, the product of a and b is generally written just by juxtaposing: ab. *Our convention is to adopt the multiplicative notation for an arbitrary group G.* This means that we write

$$ab \quad \text{for} \quad a * b, \quad a,b \in G,$$

$$1 \quad \text{for} \quad e,$$

$$a^{-1} \quad \text{for} \quad \bar{a}, \quad a \in G.$$

Actually, we may sometimes wish not to adopt the second of these conventions, preferring to retain e for the neutral element. (We also sometimes write $a \cdot b$ for ab if the meaning of the resulting symbol would thereby be clarified.)

We follow the logic of this convention in adopting the notation of exponents. Thus if n is a positive integer and $a \in G$, where G is a group, we define a^n inductively by

$$a^1 = a, \quad a^{n+1} = a^n a. \tag{1.1}$$

We extend the convention to any integer exponent by defining

$$a^0 = 1, \tag{1.2}$$

and, if n is a positive integer,

$$a^{-n} = (a^{-1})^n. \tag{1.3}$$

The group axioms (in particular, the associative law) then immediately validate the *laws of indices*:

PROPOSITION 1.4

$$a^m a^n = a^{m+n}, (a^n)^m = a^{mn}.$$

On the other hand, if a *commutative* group G is given, it is customary to adopt *additive* notations instead of *multiplicative* notations. When a commutative group is written additively, we refer to it as an *abelian* group.

Thus in an abelian group A we write

$$a+b \quad \text{for} \quad a*b, \quad a,b\in A,$$

$$0 \quad \text{for} \quad e,$$

$$-a \quad \text{for} \quad \bar{a}, \quad a\in A,$$

$$a-b \quad \text{for} \quad a+(-b), \quad a,b\in A,$$

$$na \quad \text{for} \quad a^n, \quad a\in A.$$

With the last convention, the laws of indices become

$$ma+na=(m+n)a, \quad m(na)=(mn)a. \tag{1.4}$$

Our final convention has to do with the associative law. This law asserts that the two ways of forming the product (note the multiplicative convention) of a,b, c, in that order, yield the same result:

$$(ab)c=a(bc).$$

Thus we write abc for this product, and the notation is unambiguous. Similarly, given k elements $a_1,a_2,\ldots,a_k$ in the group G, we have many (how many?) ways of forming their product in the given order, and, by virtue of the associative law, all these ways yield the same element of G. Thus we may write $a_1a_2\cdots a_k$ for this element. The reader will readily appreciate the simplification produced by this convention; he should rewrite the proof of Proposition 1.3(ii) if he needs convincing.

We end this section by establishing an important criterion for G to be a group; we will at the same time be exemplifying our conventions.

PROPOSITION 1.5. *Let* G *be a nonempty set with a given associative binary operation. Then* G *is a group if and only if, for all* a, b *in* G, *the equations*

$$ax=b, \quad ya=b \tag{1.5}$$

have solutions in G.

PROOF.   Let G be a group. Then

$$aa^{-1}b=b, \quad ba^{-1}a=b,$$

so that the equation $ax=b$ has the solution $x=a^{-1}b$, and the equation $ya=b$ has the solution $y=ba^{-1}$.

Conversely, suppose that the equations (1.5) always have solutions in G. Then, to each a in G, we find e_a such that $ae_a=a$. We show that e_a is independent of a. For if $b \in G$, there exists y in G with $ya = b$. Then $be_a = yae_a = ya = b$. Thus $e = e_a$ is a right neutral element for G. Again appealing to (1.5) we find, for each a in G, an element a^{-1} in G such that $aa^{-1} = e$. The proof is completed by appealing to Proposition 1.2.

DEFINITION 1.1.   The number (possibly infinite) of elements in a group is called the *order* of the group. The group is *finite* (*infinite*) if its order is finite (infinite).

2. SUBGROUPS

Consider the set of symmetries of the square

Plainly we may *describe* any such symmetry by listing the vertices that the symmetry maps to vertices $1, 2, 3, 4$ respectively, that is, by a *permutation* of the set $(1, 2, 3, 4)$. Moreover, the set of symmetries forms a group under composition—the same operation that provides a group structure in the set of all permutations of $(1, 2, 3, 4)$. On the other hand, not every permutation yields a symmetry of the square; for example, there is no symmetry that interchanges vertices 1 and 2, leaving 3 and 4 fixed. Thus the set of symmetries is a proper subset of the set of permutations of $(1, 2, 3, 4)$, and we would naturally like to say that the group of symmetries is a proper subgroup of the group of permutations of $(1, 2, 3, 4)$. We are thus led to the following definition.

DEFINITION 2.1.   Let $(G, *)$ be a group and let H be a subset of G that is closed under $*$, that is, $a * b \in H$ if a, $b \in H$. Then we may regard $*$ as a binary operation on H and we call $(H, *)$ a *subgroup* of $(G, *)$ if $(H, *)$ is itself a group. Notice that $(\{e\}, *)$ and $(G, *)$ are, by this definition, subgroups of $(G, *)$. We call $(H, *)$ a *proper* subgroup if H is not a singleton and $H \neq G$.

PROPOSITION 2.1.   *If* H *is a subgroup of* G *then* (*i*) *the neutral element of* H *is the neutral element of* G; (*ii*) *for all* a *in* H, *the inverse of* a *in* H *is the inverse of* a *in* G.

PROOF.
    (i) Since e is the unique idempotent of G (that is, the unique solution of $x^2 = x$) and H possesses an idempotent, it follows that $e \in H$ and is the

neutral element of H (see Exercise 1.2).

(ii) Since a^{-1} is the unique solution of $ax = e$ in G and this equation has a solution in H, it follows that $a^{-1} \in H$ and is the inverse of a in H (see Exercise 1.1).

This proposition tidies up the concept of subgroup. The next enables us to recognize a subgroup.

PROPOSITION 2.2. *Let* G *be a group and* H *a nonempty subset of* G. *Then* H *is a subgroup of* G *if and only if it is closed under* $*$ *and* $a^{-1} \in H$ *for all* a *in* H.

PROOF. The conditions are clearly necessary, in view of Proposition 2.1(ii). Now suppose the conditions fulfilled; we must show that $(H, *)$ is a group. $\mathbf{G_1}$ is satisfied, since it holds in G. As to $\mathbf{G_2}$, let $a \in H$ (note that H is not empty). Then $a^{-1} \in H$, so that $e = aa^{-1} \in H$. Finally, $\mathbf{G_3}$ holds by hypothesis.

We have two important corollaries:

COROLLARY 2.3. *Let* G *be a group and* H *a nonempty subset of* G. *Then* H *is a subgroup of* G *if and only if* $ab^{-1} \in H$ *for all* a,b *in* H.

PROOF. Suppose that H is a subgroup and $a,b \in H$. Then $b^{-1} \in H$ so that $ab^{-1} \in H$. Conversely, suppose the conditions fulfilled, and let $a \in H$. Then $e = aa^{-1} \in H$. It further follows that $a^{-1} = ea^{-1} \in H$. Now let $a,b \in H$. Then $a, b^{-1} \in H$, so that $ab = a(b^{-1})^{-1} \in H$, and H is a subgroup according to Proposition 2.2.

COROLLARY 2.4. *Let* G *be a group and* H *a finite nonempty subset of* G. *Then* H *is a subgroup of* G *if and only if it is closed under* $*$.

PROOF. The condition is trivially necessary. Now suppose the condition fulfilled and let $b \in H$. Then all the positive powers of b are in H. It cannot be that all the positive powers are distinct, since H is finite. Thus there exists a positive integer n such that $b^n = e$. (See Exercise 1.3.) In fact, there is no harm in taking $n \geqslant 2$, since if $n = 1$ then $b = e$, and thus $b^m = e$ for all positive integers m. But then $b^{n-1} = b^{-1}$ and $n - 1$ is positive, so that $b^{-1} \in H$. It thus follows that $ab^{-1} \in H$ if $a,b \in H$, so that H is a subgroup according to Corollary 2.3.

We usually apply Corollary 2.4 when G is itself finite; notice, however, that it is sufficient for our argument that G be a torsion group, that is, that every element of G be of finite order (see Exercise 1.3). On the other hand, the criterion of Corollary 2.4 is not applicable without some restriction (on H). For if we take G to be the additive group of integers and H to be the subset of positive integers, then H is closed under addition but is not a subgroup since it does not contain inverses.

From Corollary 2.3 we immediately infer the following important fact.

PROPOSITION 2.5. *Let* H_α *be a family of subgroups of* G *indexed by the set* A. *Then* $\bigcap_{\alpha \in A} H_\alpha$ *is a subgroup of* G.

PROOF. Since $e \in H_\alpha$ for all α, $e \in \bigcap_\alpha H_\alpha$, so $\bigcap_\alpha H_\alpha$ is not empty. If $a, b \in \bigcap_\alpha H_\alpha$, then $a, b \in H_\alpha$, all α, so that $ab^{-1} \in H_\alpha$, all α, whence $ab^{-1} \in \bigcap_\alpha H_\alpha$.

We may apply this proposition in the following way. Let S be a subset of the group G. We can then consider the collection $\{H_\alpha\}$ of subgroups of G which contain S. This collection is not empty, since $S \subseteq G$. Then $H_S = \bigcap_\alpha H_\alpha$ is a subgrroup of G containing S and is clearly the *smallest* subgroup of G containing S, since for any other such subgroup $H_\alpha, H_S \subseteq H_\alpha$. We call H_S the subgroup of G *generated* by S. If $H_S = G$ we call S a set of *generators* of G. A group is said to be *cyclic* if it may be generated by a single element. It is said to be *finitely generated* if it may be generated by a finite set of elements. Notice that, according to the definition given in Exercise 1.3, the *order* of an element a in G is the order of the subgroup it generates.

We may adopt a different point of view with regard to H_S which is often more useful in practice. Let us consider the collection of all elements of G of the form

$$a_1 a_2 \cdots a_n \quad \text{(variable } n) \tag{2.1}$$

where, for each i, either a_i or a_i^{-1} belongs to S; we interpret (2.1) as the element e if $n = 0$. This collection is nonempty (even if S is empty it contains e), and it is plain from the criterion of Corollary 2.3 that it is a subgroup of G. Moreover, if H is any subgroup of G containing S, then H must contain all the elements (2.1). Thus this collection is, in fact, H_S. We may say that H_S is the set of all words formed from the set of symbols $\{a\}$, $a \in S$, and their inverses.

We close this section with an important example.

EXAMPLE 2.6. We study the subgroups of **Z**. Of course, we have the trivial (improper) subgroup $\{0\}$. We also have the subgroups $m\mathbf{Z}$, where m is a positive integer. We show that these are the only subgroups of **Z**. Thus let H be a subgroup of **Z**, $H \neq \{0\}$. Then H contains a nonzero integer and, being a subgroup, must therefore contain a positive integer (if $n \in H$, $(-n) \in H$). Let m be the smallest positive integer in H. Then $m\mathbf{Z} \subseteq H$ [see (1.4)]. We show that $H \subseteq m\mathbf{Z}$. Thus let $h \in H$. By the Euclidean algorithm, there exist q, r such that

$$h = mq + r, \quad 0 \leqslant r < m. \tag{2.2}$$

Since $h, m \in H$ and $r = h - mq$, it follows that $r \in H$. Since m is the smallest positive integer in H, it follows from (2.2) that $r = 0$. Thus $h = mq, h \in m\mathbf{Z}, H \subseteq m\mathbf{Z}, H = m\mathbf{Z}$.

3. COSETS, LAGRANGE'S THEOREM, AND NORMAL SUBGROUPS

Let G be a group and H a subgroup of G. If $a \in G$ we write aH for the subset of G consisting of elements $ah, h \in H$; we call aH a *right coset* of H in G. Similarly we may speak of *left cosets Ha*.

We introduce the relation $\sim_H$ on G by declaring $a \sim_H b$ if $ab^{-1} \in H$.

PROPOSITION 3.1. *The relation $\sim_H$ is an equivalence relation on* G.

PROOF. $\sim_H$ is reflexive because $1 \in H$, $\sim_H$ is symmetric because H contains inverses, and $\sim_H$ is transitive because H is closed under $*$. (Indeed the reader will observe that $\sim_S$ may be defined for any subset S of G and $\sim_S$ is an equivalence relation if and only if S is a subgroup of G.)

We now note that the equivalence classes with respect to $\sim_H$ are precisely the left cosets Ha. Thus two cosets Ha_1, Ha_2 are either disjoint or identical. Let us now write $|S|$ for the *order* of S, that is, the number of elements in S.

LEMMA 3.2.

$$|Ha_1| = |Ha_2| = |a_1 H| = |a_2 H|.$$

PROOF. It is plainly sufficient to prove that $|Ha| = |H|$, for then a similar proof shows that $|aH| = |H|$. But the function $h \rightarrow ha$ is a one-one correspondence between the sets H and Ha.

We are now able to prove a basic theorem of group theory.

THEOREM 3.3 (LAGRANGE'S THEOREM). *Let* H *be a subgroup of the finite group* G. *Then* |H| *divides* |G|.

PROOF. Partition the set G into disjoint left cosets $Ha_1, Ha_2, \ldots, Ha_n$. Then $|Ha_1| = |Ha_2| = \cdots = |Ha_n| = |H|$, by Lemma 3.2. Thus

$$|G| = n|H|. \tag{3.1}$$

The number n appearing in this proof is called the *index* of H in G. It is plain that the index is also the number of distinct *right* cosets of H in G.

We may plainly extend (3.1) to the case of an arbitrary group G, when (3.1) becomes a relation between cardinal numbers. It is then not immediate that the number of left cosets of H in G equals the number of right cosets of H in G; but this is true and follows from the fact, whose proof we omit, that the function

$$aH \mapsto Ha^{-1}$$

sets up a one-one correspondence between right cosets and left cosets.

As a consequence of Theorem 3.3, we have

COROLLARY 3.4. *Let* G *be a finite group and* $a \in G$. *Then the order of* a *divides* $|G|$.

We return to the situation of a subgroup H of a group G. A particularly important case is that in which every left coset is also a right coset. Now if $Ha = bH$, then, since $a \in Ha$, we have $a \in bH$, so $bH = aH$. Thus we may describe this situation by saying that $Ha = aH$ for all a in G.

DEFINITION 3.1. The subgroup H of G is called *normal* or *self-conjugate* if $Ha = aH$ for all a in G.

EXAMPLES 3.5.

1. If G is commutative, every subgroup is normal. (This shows that the concept of a normal subgroup is a natural generalization to noncommutative groups of the notion of subgroup of a commutative group).

2. Let G be a group and let $C \subseteq G$ be the subset of G consisting of all $a \in G$ such that $ab = ba$ for all b in G. The reader will readily verify that C is a normal subgroup of G. It is called the *center* of G.

3. Consider P_3, the group of all permutations of the three symbols $1, 2, 3$. The elements of P_3, expressed by means of the values of the permutation on $1, 2, 3$ respectively, are

$$a_1 = (1, 2, 3)$$
$$a_2 = (1, 3, 2)$$
$$a_3 = (2, 1, 3)$$
$$a_4 = (2, 3, 1)$$
$$a_5 = (3, 1, 2)$$
$$a_6 = (3, 2, 1)$$

The *group table* is

P_3	a_1	a_2	a_3	a_4	a_5	a_6
a_1	a_1	a_2	a_3	a_4	a_5	a_6
a_2	a_2	a_1	a_4	a_3	a_6	a_5
a_3	a_3	a_5	a_1	a_6	a_2	a_4
a_4	a_4	a_6	a_2	a_5	a_1	a_3
a_5	a_5	a_3	a_6	a_1	a_4	a_2
a_6	a_6	a_4	a_5	a_2	a_3	a_1

Here the entry in row a_i, column a_j, is the product $a_i a_j$ in the group P_3, that is, the permutation that consists of first executing a_i, then a_j. Of course, a_1 is the neutral element[†] of P_3.

Consider the subset $A = \{a_1, a_2\}$ of P_3. It is clear from the table that this is a subgroup. Then the left cosets of A in P_3 are

$$A = \{a_1, a_2\}, \quad Aa_3 = \{a_3, a_4\}, \quad Aa_5 = \{a_5, a_6\},$$

while the right cosets of A in P_3 are

$$A = \{a_1, a_2\}, \quad a_3 A = \{a_3, a_5\}, \quad a_4 A = \{a_4, a_6\}.$$

Thus we see that A is not normal in G.

Now consider the subset $B = \{a_1, a_4, a_5\}$ of P_3. Again, it is clear from the table that this is a subgroup. Then the left cosets of B in P_3 are

$$B = \{a_1, a_4, a_5\}, \quad Ba_2 = \{a_2, a_6, a_3\},$$

while the right cosets of B in P_3 are

$$B = \{a_1, a_4, a_5\}, \quad a_2 B = \{a_2, a_3, a_6\}.$$

Thus B is normal in G.

This last example illustrates a general fact.

PROPOSITION 3.6. *Let* H *be a subgroup of* G *of index 2. Then* H *is normal in* G.

PROOF. Choose $a \in G$, $a \notin H$. Then $G = H \cup aH = H \cup Ha$, where both unions are disjoint. Thus $aH = Ha$, each being the complement of H in G.

We next explain the use of the term "self-conjugate" for a normal subgroup. Given any subgroup H of G, and any element $a \in G$, we may form the subset $a^{-1}Ha$ of G, consisting of all elements of G of the form $a^{-1}ha$, $h \in H$. It is easy to see that $a^{-1}Ha$ is also a subgroup of G, and it is said to be *conjugate* to H (indeed, the element $a^{-1}ha$ is also said to be conjugate to h). We also say that $a^{-1}Ha$ is obtained from H by *conjugating* by the element a. Then the reader will easily prove

PROPOSITION 3.7. *The subgroup* H *of* G *is normal if and only if it coincides with all its conjugates.*

Given a subgroup H of G, we may form the set G/H of right cosets of

---

[†]It is customary to express elements of P_n as products of cyclic permutations operating on mutually disjoint subsets of the set $1,2,3,\ldots,n$. In this notation we have $a_1 = e, a_2 = (23), a_3 = (12), a_4 = (123), a_5 = (132), a_6 = (13)$.

H in G. This set is called the (right) *coset space* of H in G. We could, of course, also form the left coset space. However, we now assume H *normal* so that the distinction between right and left disappears. In this case we are able to endow G/H with a group structure in a very natural way, namely, by setting

$$aH \cdot bH = abH. \tag{3.2}$$

However, this definition contains a concealed subtlety. For what it says is this: "Take two cosets, pick elements in those cosets, multiply the elements, and find the coset in which the product lies." To justify this procedure we have to ensure that the resulting coset depends only on the cosets from which we started, and not on the particular elements we picked. (We call such chosen elements *representatives*.)

We will see how the justification depends on the assumption that H is normal. For let $a' \in aH$, $b' \in bH$. Then $a' = ah_1$, $b' = bh_2$, $h_1, h_2 \in H$, so that

$$a'b' = ah_1bh_2 = ab(b^{-1}h_1b)h_2.$$

Now, since H is normal, $b^{-1}h_1b \in H$, so that $(b^{-1}h_1b)h_2 \in H$. Thus $a'b' = abh$, $h \in H$, so that $a'b' \in abH$, and we have proved that the coset on the right of (3.2) depends only on the cosets aH, bH and not on the particular representatives a,b. Thus (3.2) is a valid definition.

On the other hand, if H is not normal, (3.2) is invalid. Consider the example $G = P_3$, $H = A$ of Example 3.5.3. Then if we multiply an element in the right coset A with an element in the right coset a_3A we have the possibilities

$$a_1a_3 = a_3, \quad a_1a_5 = a_5, \quad a_2a_3 = a_4, \quad a_2a_5 = a_6.$$

Thus we may land in the coset a_3A or the coset a_4A, and (3.2) is an ambiguous rule.

Now, when H is normal, (3.2) defines a binary operation in G/H, and it is easy to see that it yields a group structure in G/H. For associativity follows immediately from associativity in G; the neutral element is the coset H itself; and the inverse of aH is $a^{-1}H$. We are thus led to the following important definition.

DEFINITION 3.2. If H is a normal subgroup of the group G, then the coset space G/H, endowed with the group structure given by (3.2), is called the *factor group* or *quotient group* of G by H.

EXAMPLE. 3.8. Let $G = \mathbf{Z}$, $H = m\mathbf{Z}$. Since G is abelian, H is normal and

$$G/H = \mathbf{Z}_m,$$

the group of residue classes modulo m (see Example 5 of Section 1). The last example illustrates the obvious fact that G/H is commutative if G is commutative.

4. HOMOMORPHISMS

Since groups are sets with additional structure, it is natural to single out those functions between groups that preserve the structure. It is a matter of taste whether we regard the structure in the group G as specified by the binary operation or incorporate the neutral element and inverses into the description of the structure. It turns out that we arrive at the same notion of distinguished function between groups, or *homomorphism*, whichever viewpoint we take.

DEFINITION 4.1.   Let G, K be groups and let $\phi : G \rightarrow K$ be a function between their underlying sets. Then ϕ is a *homomorphism* if

$$\phi(g_1 g_2) = \phi(g_1)\phi(g_2),$$

for all g_1, g_2 in G. Notice that the product $g_1 g_2$ is taken in G, whereas the product $\phi(g_1)\phi(g_2)$ is taken in K. Notice also that there is always at least one homomorphism from G to K: namely, that function which sends the whole of G to the neutral element of K. We call this the *zero* or *constant* homomorphism, and write it as $0 : G \rightarrow K$.

We proceed immediately to justify our opening remarks.

PROPOSITION 4.1.   *If* $\phi : G \rightarrow K$ *is a homomorphism, then*

(i) $\phi(e_G) = e_K$;

(ii) $\phi(g)^{-1} = \phi(g^{-1}).$

(We have written e_G, e_K for the neutral elements of G, K for the sake of emphasis; we will later write $\phi 1 = 1$.)

PROOF.

(i) Let $\phi(e_G) = k$. Then $e_G^2 = e_G$, so $k^2 = \phi(e_G)\phi(e_G) = \phi(e_G^2) = \phi(e_G) = k$. Since e_K is the only idempotent in K, we have $k = e_K$.

(ii) $e_K = \phi(e_G) = \phi(gg^{-1}) = \phi(g)\phi(g^{-1})$. Thus $\phi(g^{-1}) = \phi(g)^{-1}$.

We next prove that the notion of homomorphism has two crucial "universal" properties.

PROPOSITION 4.2.

(i) *If* G *is a group, the identity function* $I : G \rightarrow G$ *is a homomorphism.*

(ii) *If* $\phi: G \rightarrow K$, $\psi: K \rightarrow L$ *are homomorphisms, so is their composite* $\psi\phi: G \rightarrow L$.

PROOF.

(i) The proof is obvious.

(ii) $\psi\phi(g_1 g_2) = \psi(\phi(g_1)\phi(g_2)) = \psi\phi(g_1)\psi\phi(g_2)$.

Since composition of functions is always associative, Proposition 4.2 may be summed up by saying that the collection of groups and homomorphisms constitutes a *category* (with respect to composition of functions). Now, any category carries with it the notions of *isomorphism* and *isomorphic objects* (see Chapter 3, Section 1), so that we are led to the following definition, where we write 1 for the identity homomorphism I.

DEFINITION 4.2. A homomorphism $\phi: G \rightarrow K$ is called an *isomorphism* if there exists a homomorphism $\psi: K \rightarrow G$ such that $\psi\phi = 1_G$, $\phi\psi = 1_K$. Then ψ is also an isomorphism, which is clearly determined by ϕ, and is called the *inverse* of ϕ. We also say that G and K are *isomorphic*, and write $G \cong K$, or $\phi: G \cong K$.

It is a trivial (and purely categorical) observation that the relation $\cong$ is an equivalence relation. However, we have the following important (and noncategorical!) criterion for isomorphism.

THEOREM 4.3. *A homomorphism $\phi: G \rightarrow K$ is an isomorphism if and only if it is surjective and injective (as a function).*

PROOF. An isomorphism is certainly a one-one correspondence of the underlying sets. Conversely, let $\phi: G \rightarrow K$ be a homomorphism which is also a one-one correspondence of the underlying sets. There is then certainly an inverse function $\psi: K \rightarrow G$, and it only remains to show that ψ is a homomorphism. Now

$$\phi\psi(k_1 k_2) = k_1 k_2 = \phi\psi(k_1)\phi\psi(k_2) = \phi(\psi(k_1)\psi(k_2)).$$

Since ϕ is injective it follows that $\psi(k_1 k_2) = \psi(k_1)\psi(k_2)$, so the theorem is proved.

REMARK. It has been customary to *define* an isomorphism of groups as a homomorphism that is surjective and injective. This custom is regrettable,[†] since the notion of isomorphic objects in a category is of the utmost generality. Thus the custom leads the student into error when he comes to study other structures, that is, other categories. As already pointed out, our definition of isomorphism is valid in any category (a homomorphism is just

---

[†] Of course, the custom springs from a similar custom in the category of sets and functions —which is certainly just as regrettable!

a *morphism* of the category of groups); but it is false that every (continuous) homomorphism of topological groups that is surjective and injective is an isomorphism.

We now simplify the test for isomorphism further by simplifying the test for injectivity.

PROPOSITION 4.4. *A homomorphism* $\phi : G \to K$ *is injective if and only if* $\phi^{-1}(e) = e$.

PROOF. Since $\phi(e) = e$, the condition is clearly necessary. To prove it sufficient, let $\phi^{-1}(e) = e$, and let $\phi(g_1) = \phi(g_2)$. Then

$$\phi(g_1 g_2^{-1}) = \phi(g_1) \phi(g_2)^{-1} = e,$$

so that $g_1 g_2^{-1} = e$, and $g_1 = g_2$.

We may paraphrase this proposition by asserting that, in testing for injectivity, we may "localize" the test at e.

PROPOSITION 4.5. *Let* $\phi : G \to K$ *be a homomorphism. Then*

(i) *For any subgroup* H *of* K, *the image* $\phi(H)$ *is a subgroup of* K.

(ii) *For any subgroup* L *of* K, *the counterimage* $\phi^{-1}(L)$ *is a subgroup of* G, *which is normal if* L *is normal.*

PROOF.

(i) $\phi(H)$ is not empty, since H is not empty. Because $\phi(ab^{-1}) = \phi(a)$ $\phi(b)^{-1}$, it follows from Corollary 2.4 that $\phi(H)$ is a subgroup of K.

(ii) $\phi^{-1}(L)$ is not empty, since $\phi(e) = e$. If $c, d \in \phi^{-1}(L)$, then $\phi(cd^{-1})$ $= \phi(c)\phi(d)^{-1}$, so $\phi(cd^{-1}) \in L$, $cd^{-1} \in \phi^{-1}(L)$ and $\phi^{-1}(L)$ is a subgroup of G. If L is a normal subgroup, then for $c \in \phi^{-1}(L)$, $a \in G$,

$$\phi(a^{-1}ca) = \phi(a)^{-1}\phi(c)\phi(a) \in L,$$

so that $\phi^{-1}(L)$ is normal in G (see Exercise 3.1).

We call $\phi(G)$ the *image* of ϕ, so that the image of ϕ is a subgroup of K. Since $\{e\}$ is a normal subgroup of K, $\phi^{-1}(e)$ is a normal subgroup of G; we call it the *kernel* of ϕ. We have thus seen that the kernel of a homomorphism with domain G is a normal subgroup of G; however, the converse is also true. If we let H be a normal subgroup of G, we have constructed (Definition 3.2) the factor group G/H, and there is plainly a canonical homomorphism

$$\pi : G \to G/H$$

given by

$$\pi(g) = gH, \quad g \in G. \tag{4.1}$$

The following proposition is obvious.

PROPOSITION 4.6.   *The homomorphism π is surjective with kernel* H.

Our next two results characterize (i) the kernel of a homomorphism and (ii) the factor group by means of a "universal" property. We do not define precisely what is meant by this term (see [6]), but the reader should suspect a categorical flavor to the property, and will find it much referred to in Chapter 3.

THEOREM 4.7.   *Let* H *be the kernel of* $\phi: G \to K$ *and let* $\iota: H \to G$ *be the embedding, so that* ι *is an injective homomorphism. Then* $\phi\iota = 0$. *Moreover, if* $\lambda: L \to G$ *is a homomorphism such that* $\phi\lambda = 0$, *then there exists a unique homomorphism* $\bar{\lambda}: L \to H$ *such that* $\iota\bar{\lambda} = \lambda$.

Conversely, given $\phi: G \to K$, *let* $\iota': H' \to G$ *be a homomorphism such that* $\phi\iota' = 0$ *and such that, if* $\lambda: L \to G$ *satisfies* $\phi\lambda = 0$, *then there exists a unique homomorphism* $\bar{\lambda}'$ *with* $\iota'\bar{\lambda}' = \lambda$. *Then* ι' *is injective and there exists a unique isomorphism* $\theta: H \to H'$ *such that* $\iota'\theta = \iota$.

We may express the first part of the enunciation by means of the commutative diagram (4.2). Here (and henceforth) $\rightarrowtail$ denotes an injective homomorphism, and a dotted arrow with a label denotes a homomorphism whose existence and uniqueness are asserted or have been proved. The second part of the statement effectively claims that diagram (4.2) characterizes the kernel or, rather, the embedding ι of the kernel.

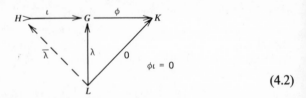

$$\phi\iota = 0$$

(4.2)

PROOF OF THEOREM 4.7.   That the kernel enjoys property (4.2) is obvious; for if $\phi\lambda = 0$, then λ maps L to the kernel of ϕ. Thus we concentrate on the converse.

We first note that, by the property (4.2) of ι, it follows, since $\phi\iota' = 0$, that there exists a unique $\theta': H' \to H$ such that $\iota\theta' = \iota'$. However, by the property (4.2) of ι', it follows, since $\phi\iota = 0$, that there exists a unique $\theta: H \to H'$ such that $\iota'\theta = \iota$. But then

$$\iota\theta'\theta = \iota, \quad \iota'\theta\theta' = \iota'. \tag{4.3}$$

Now the *uniqueness* part of property (4.2) asserts that $\bar{\lambda}$ is determined by $\iota\bar{\lambda}$ (this is, in fact, equivalent to the statement that ι is injective, as the reader

should prove). Thus from (4.3) we infer

$$\theta'\theta = 1_H, \quad \theta\theta' = 1_{H'},$$

so that θ is an isomorphism. Of course it now follows that ι' is injective, but, as we have said, this is already, and fairly trivially, implied by the statement, in the enunciation of the theorem, that $\bar{\lambda}'$ is unique.

We now characterize the factor group in similar fashion.

THEOREM 4.8. *Let* H *be a normal subgroup of* G *with* $\iota : H \to G$ *the embedding and* $\pi : G \to G/H$ *the projection onto the factor group. Then* $\pi\iota = 0$. *Moreover, if* $\lambda : G \to L$ *is a homomorphism such that* $\lambda\iota = 0$, *then there exists a unique homomorphism* $\bar{\lambda} : G/H \to L$ *such that* $\bar{\lambda}\pi = \lambda$.

Conversely, let $\pi' : G \to Q$ *be a homomorphism such that* $\pi'\iota = 0$ *and such that, if* $\lambda : G \to L$ *satisfies* $\lambda\iota = 0$, *then there exists a unique homomorphism* $\bar{\lambda}' : Q \to L$ *with* $\bar{\lambda}'\pi' = \lambda$. *Then* π' *is surjective and there exists a unique isomorphism* $\theta : G/H \to Q$ *with* $\theta\pi = \pi'$.

We may express the first part of the enunciation [compare (4.2)] by means of the commutative diagram (4.4) where (and henceforth) $\to\!\!\!\to$ denotes a surjective homomorphism. Again, the second part of the statement effectively claims that diagram (4.4) characterizes the factor group or, rather, the projection π onto the factor group.

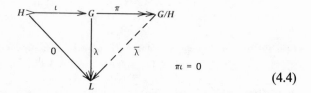

$$\pi\iota = 0 \tag{4.4}$$

PROOF OF THEOREM 4.8. Given $\lambda : G \to L$, set $\bar{\lambda}(gH) = \lambda(g)$. Then $\bar{\lambda}$ is well defined since $\lambda(H) = \{e\}$. Moreover $\bar{\lambda}$ is evidently a homomorphism and $\bar{\lambda}\pi = \lambda$. The last equation obviously determines $\bar{\lambda}$ since π is surjective.

The converse is proved just as for Theorem 4.7. Since $\pi'\iota = 0$, property (4.4) for π yields a unique $\theta : G/H \to Q$ with $\theta\pi = \pi'$. But then property (4.4) for π' yields a unique $\theta' : Q \to G/H$ with $\theta'\pi' = \pi$. From

$$\theta'\theta\pi = \pi, \quad \theta\theta'\pi' = \pi',$$

a further application of the uniqueness part of property (4.4) yields

$$\theta'\theta = 1, \quad \theta\theta' = 1,$$

so that θ is an isomorphism. It then follows, of course, that π' is surjective; however, this again follows, but not so trivially, directly from the uniqueness of $\bar{\lambda}'$ in the enunciation. We revert to this point later (see Exercise 4.6 and Chapter 3).

REMARKS.
(i) In Theorem 4.7 we proved that, given $\phi: G \to K$, any ι satisfying property (4.2) is essentially unique. Thus the construction of the kernel of ϕ yields an *existence* proof of such an ι and Theorem 4.7 attests its *uniqueness*. Similarly the construction of the factor group yields an existence proof of a homomorphism π satisfying property (4.4) and Theorem 4.8 attests its uniqueness. We find this to be typical of categorical reasoning — special constructions, within a given category, are required to establish the existence of the object described by some universal property, but very general categorical arguments establish its uniqueness.

(ii) There is an evident difference between (4.2) and (4.4) in that (4.2) relates to an arbitrary homomorphism ϕ, whereas (4.4) relates to the embedding ι of a normal subgroup. If $\phi: K \to G$ is a homomorphism such that $\phi(K)$ is normal in G, we may replace (4.4) by diagram (4.5), and the argument goes through as before.

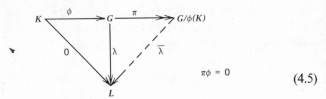

$$\pi\phi = 0 \qquad (4.5)$$

For the general case, we need to introduce a new concept. We first note the following elementary fact (see Exercise 3.1).

PROPOSITION 4.9.   *The intersection of a family of normal subgroups is again a normal subgroup.*

From this we infer that, given any subgroup H of G, there exists a smallest normal subgroup of G containing H. We call this the *normal hull* of H and write it $\hat{H}$. Then the general case is expressed in diagram (4.6). For, since the kernel of any homomorphism is a normal subgroup, it is clear that if λ annihilates $\phi(K)$ it must annihilate $\widehat{\phi(K)}$. We call $G/\widehat{\phi(K)}$, or even the projection π (4.6), the *cokernel* of ϕ.

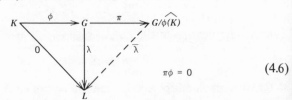

$$\pi\phi = 0 \qquad (4.6)$$

Let $\phi: G \to K$ have kernel H. An immediate application of property (4.4), or a direct argument, yields $\bar\phi: G/H \to K$ such that $\bar\phi\pi = \phi$. Then $\bar\phi$ is injective. To prove this we have to show that if $\bar\phi(gH) = e$ then $g \in H$ (see Proposition 4.4). But $\bar\phi(gH) = \phi(g)$, so the conclusion is obvious.

THEOREM 4.10. *If $\phi: G \to K$ has kernel* H, *then ϕ induces*

$$\phi_1: G/H \cong \phi G.$$

For $\bar\phi$ maps G/H onto ϕG, so that ϕ_1 is surjective and injective. Here ϕ_1 and $\bar\phi$ differ only in the specification of the range.

A second isomorphism theorem is reminiscent of, and related to, a property of rational fractions.

THEOREM 4.11. *Suppose that* $H \subseteq K \subseteq G$ *with* H, K *normal in* G. *Then* K/H *is normal in* G/H *and*

$$(G/H)/(K/H) \cong G/K.$$

PROOF. The function $gH \mapsto gK$ plainly induces a surjective homomorphism $\rho: G/H \twoheadrightarrow G/K$. We study the kernel of ρ. Now there is an obvious embedding of K/H in G/H and its image consists just of those cosets gH with $g \in K$. But these are precisely the elements of G/H in the kernel of ρ. Thus K/H is the kernel of ρ, hence normal in G/H, and the proof is completed by applying Theorem 4.10.

5. DIRECT AND FREE PRODUCTS

In this section we introduce two fundamental ways in which we may construct a single group from a family of groups. As in the case of kernels and cokernels in Section 4, we then characterize these constructions by universal properties, so that the constructions themselves appear as existence theorems for groups with the given properties.

Let I be an indexing set and let $\{G_i\}$, $i \in I$, be a family of groups indexed by I. We construct a group G as follows. The underlying set of G is the Cartesian product $\Pi_{i \in I} G_i$. The group operation in G is defined componentwise, that is,

$$\{ g_i \} \{ g_i' \} = \{ g_i g_i' \}. \tag{5.1}$$

It is plain that (5.1) gives us a group structure in G. The group so defined is called the *direct product* of the groups G_i, and we write

$$G = \prod_{i \in I} G_i.$$

Each G_i is called a *direct factor* in G. Notice that, with this definition, the projections $p_i : G \rightarrow G_i$ become homomorphisms; indeed, (5.1) is the unique group structure such that each p_i is a homomorphism, and we will regard the p_i as part of the structure of the direct product, so that the direct product of the groups G_i is strictly the collection $(G; p_i)$. We sometimes call this the *external* direct product, for reasons that appear later in this section.

Now let X be a group and let $\xi_i : X \rightarrow G_i$ be homomorphisms, $i \in I$. It is then obvious that there is a unique homomorphism $\xi : X \rightarrow G$ such that $p_i \xi = \xi_i$, $i \in I$, namely,

$$\xi x = \{\xi_i x\}, \quad x \in X, \quad i \in I. \tag{5.2}$$

We refer to this as *property D* of $(G; p_i)$, and write $\xi = \{\xi_i\}$. This is the universal property which characterizes $(G; p_i)$. Precisely we have

THEOREM 5.1. *Given a group* G' *and homomorphisms* $p_i' : G' \rightarrow G_i$, $i \in I$, *suppose that* $(G'; p_i')$ *has property* D, *that is, suppose that, given* $\xi_i : X \rightarrow G_i$, *there exists a unique homomorphism* $\xi' : X \rightarrow G'$ *with* $p_i' \xi' = \xi_i$, $i \in I$. *Then there exists a unique isomorphism* $\theta : G \rightarrow G'$ *such that* $p_i' \theta = p_i$, $i \in I$.

We leave the proof to the reader (compare Theorem 4.7).

Now there are embeddings $\iota_j : G_j \rightarrow G$, given by

$$\iota_j(g_j) = \{g_i'\},$$

where

$$g_i' = e, \quad i \neq j,$$

$$g_j' = g_j.$$

Indeed (of course) these embeddings are guaranteed, and specified, by property D; we merely have to take $X = G_j, \xi_i = 0, i \neq j, \xi_j = 1$. We note that G_j is embedded in G as a *normal* subgroup.

It is now natural to raise the following question. Suppose given a group H and a family G_i, $i \in I$, of subgroups of H. Does there exist an isomorphism $\omega : H \rightarrow \prod_i G_i$ such that $\omega | G_i = \iota_i$, $i \in I$? We will study this question only when the indexing set I is *finite*. It is clear that when such an ω exists, then the study of the structure of H reduces to that of its *factors* G_i. If ω exists, we say that H is the *internal* direct product of its subgroups G_i.

We introduce the notation $\square \, G_i$ for the smallest subgroup of H containing the groups G_i, $i \in I$.

THEOREM 5.2. *Let* $G_1, G_2, \ldots, G_k$ *be subgroups of* H. *Then* H *is the internal direct product of these subgroups if and only if the following three* *conditions hold*:

   (*i*) *Each* G_i *is normal in* H,

   (*ii*) $H = \square_{i=1}^{k} G_i$,

   (*iii*) $G_j \cap \square_{i \neq j} G_i = \{e\}, j = 1, 2, \ldots, k$.

PROOF. Since the equivalents of (i), (ii), (iii) are true statements of the subgroups $\iota_1(G_1), \ldots, \iota_k(G_k)$ of $G = \Pi G_i$, they are necessary conditions for H to be the internal direct product of $G_1, \ldots, G_k$.

To prove them sufficient, we remark that they imply that every element of H is uniquely expressible as

$$h = g_1 g_2 \cdots g_k, \quad g_i \in G_i, \tag{5.3}$$

and that

$$(g_1 g_2 \cdots g_k)(g_1' g_2' \cdots g_k') = g_1 g_1' g_2 g_2' \cdots g_k g_k'. \tag{5.4}$$

For (i) and (ii) imply that every $h \in H$ has such an expression (5.3) (see Exercise 3.3). To prove uniqueness, suppose that

$$x_1 x_2 \cdots x_k = y_1 y_2 \cdots y_k, \quad x_i, y_i \in G_i, \quad i = 1, 2, \ldots, k.$$

Thus

$$y_1^{-1} x_1 = (y_2 \cdots y_k)(x_2 \cdots x_k)^{-1}.$$

But the left-hand side belongs to G_1 and the right-hand side belongs to $\square_{i \neq 1} G_i$. Thus, by (iii),

$$x_1 = y_1, \quad x_2 \ldots x_k = y_2 \ldots y_k.$$

We proceed in this way to show that $x_1 = y_1, x_2 = y_2, \ldots, x_k = y_k$, establishing uniqueness. Then (5.4) follows from Exercise 3.4.

The proof of the theorem is now immediate. The uniqueness of (5.3) establishes a bijective function $\omega : H \to \Pi G_i$, given by

$$\omega(g_1 g_2 \cdots g_k) = \{ g_1, g_2, \ldots, g_k \},$$

and (5.4) asserts that ω is a homomorphism.

REMARK. For an arbitrary index set I, there is an important normal subgroup of ΠG_i, namely, the subgroup generated by the groups $\iota_i(G_i)$. This is called the *restricted* direct product of the groups G_i, and consists of those elements $\{ g_i \}$ of ΠG_i, such that $g_i = e$ for all but a finite number of values of the index i. It is plain that Theorem 5.2 generalizes to the case of an arbitrary index set I to yield necessary and sufficient conditions for H

to be the restricted internal direct product of its subgroups G_i.

When the index set I is finite there is, of course, no distinction between direct product and restricted direct product. In this case we may write, with $I = (1, 2, \ldots, k)$,

$$G = G_1 \times G_2 \times \cdots \times G_k,$$

and an element of G may be written (as we have already done!) as

$$g = (g_1, g_2, \ldots, g_k).$$

We also write

$$\xi = \{\xi_1, \xi_2, \ldots, \xi_k\}, \quad \xi_i : X \to G_i.$$

We turn now to the *free product* of a family of groups G_i, $i \in I$. We want to construct a group $\tilde{G}$, together with homomorphisms $q_i : G_i \to \tilde{G}$ such that, given any group X and homomorphisms $\xi_i : G_i \to X$, there exists a unique homomorphism $\xi : \tilde{G} \to X$ with $\xi q_i = \xi_i$. We call this *property F* of the collection $(\tilde{G}; q_i)$ and it is again easy to prove that property F characterizes $(\tilde{G}; q_i)$ up to unique (canonical) isomorphism.

The question remains whether $(\tilde{G}; q_i)$ exists. We had a very simple proof of the existence of $(G; p_i)$ verifying property D, but the question of the existence of $(\tilde{G}; q_i)$ is far subtler. We will not give all details.

We assume, conventionally, that the groups G_i are all disjoint. Then we take the underlying set of $\tilde{G}$ to consist of all "words" of the form

$$a_1 a_2 \cdots a_m, \quad m \geqslant 0, \tag{5.5}$$

where each a_j belongs to some G_i but adjacent a's do not belong to the same G_i, and $a_j \neq e$. If $m = 0$, we understand (5.5) to stand for the empty word, which we designate by 1. We call m the length of (5.5) and define the product of two words $a_1 a_2 \cdots a_m$ and $b_1 b_2 \cdots b_n$ by induction on $m + n$. Plainly we will set $11 = 1$ and $a1 = a$, $1b = b$. If the product of words whose combined length is $< m + n$ has been defined, we set

$$(a_1 a_2 \cdots a_m)(b_1 b_2 \cdots b_n)$$

$$= a_1 a_2 \cdots a_m b_1 b_2 \cdots b_n \qquad \text{if } a_m, b_1 \text{ belong to different } G_i's,$$

$$= a_1 a_2 \cdots a_{m-1}(a_m b_1) b_2 \qquad \text{if } a_m, b_1 \in G_i, a_m b_1 \neq e,$$

$$= (a_1 a_2 \cdots a_{m-1})(b_2 \cdots b_n) \qquad \text{if } a_m, b_1 \in G_i, a_m b_1 = e.$$

We omit the proof that this operation is associative; plainly 1 is the neutral element and

$$(a_1 a_2 \cdots a_m)^{-1} = a_m^{-1} \cdots a_2^{-1} a_1^{-1}.$$

Thus $\tilde{G}$ is a group and $q_i : G_i \to \tilde{G}$ is given by $q_i(a) = a$, $a \in G_i, a \neq e$, $q_i(e) = 1$. It is now easy to see that $(\tilde{G}; q_i)$ enjoys property F. For, given $\xi_i : G_i \to X$, the only choice for $\xi : G \to X$ satisfying $\xi q_i = \xi_i$ is given by $\xi(a_1 a_2 \cdots a_m) = \xi_{i_1}(a_1)\xi_{i_2}(a_2) \cdots \xi_{i_m}(a_m)$, where $a_j \in G_{i_j}$, $j = 1, 2, \ldots, m$. Moreover ξ, so defined, is indeed a homomorphism. Thus the existence of the free product is, in principle, proved. We write

$$\tilde{G} = \coprod_{i \in I} G_i,$$

or, if I is finite, $I = (1, 2, \ldots, k)$,

$$\tilde{G} = G_1 \amalg G_2 \amalg \cdots \amalg G_k.$$

We also write

$$\xi = \langle \xi_1, \xi_2, \ldots, \xi_k \rangle, \quad \xi_i : G_i \to X.$$

Finally we note that there is a canonical homomorphism $\omega : \coprod_{i \in I} G_i \to \prod_{i \in I} G_i$. We do not exhibit ω explicitly but use instead properties D and F. In fact we first construct a homomorphism under more general conditions. Suppose that we are given groups and homomorphisms $G_i, H_j, \phi_{ij} : G_i \to H_j, i \in I, j \in J$. We wish to describe a homomorphism $\phi : \coprod_{i \in I} G_i \to \prod_{j \in J} H_j$. By property F, to describe ϕ it is sufficient to give its components $\phi_i : G_i \to \prod_{j \in J} H_j$. But, by property D, to describe ϕ_i it is sufficient to give its components $(\phi_i)_j : G_i \to H_j$. We now simply take $(\phi_i)_j = \phi_{ij}$.

Consider now the special case $I = J, \phi_{ii} = 1, \phi_{ij} = 0, i \neq j$. We obtain $\omega : \coprod_{i \in I} G_i \to \prod_{i \in I} G_i$. The reader should describe ω explicitly and observe that the image of ω is precisely the restricted direct product. This observation enables us to characterize the restricted direct product by means of universal properties.

6. FREE GROUPS AND PRESENTATIONS

Let S be a set. By the *free group on the set* S we understand a group $F = F_S$ and a function $j : S \to F_S$ enjoying the following universal property: given any group G and function $f : S \to G$ there exists a unique homomorphism $\phi : F \to G$ with $\phi j = f$.

The reader will easily prove

THEOREM 6.1. *The free group on the set* S *is unique up to canonical isomorphism.*

It remains to establish the existence of the free group on S. However, this is readily achieved using the results of Section 5. For let **Z** be a cyclic infinite group, generated by a, and let $\mathbf{Z}^{(S)}$ be the free product of copies of $\mathbf{Z}$, indexed by S. Let $j: S \rightarrow \mathbf{Z}^{(S)}$ be given by

$$j(s) = q_s(a), \quad s \in S.$$

THEOREM 6.2.   *The pair* $(\mathbf{Z}^{(S)}, j)$ *is the free group on the set* S.

PROOF.   First it is plain that, given any group G and element g in G, there is a unique homomorphism $\phi_g: \mathbf{Z} \rightarrow G$, given by $\phi_g(a) = g$.

Now, given $f: S \rightarrow G$, property F guarantees us a unique homomorphism $\phi: \mathbf{Z}^{(S)} \rightarrow G$ such that $\phi q_s = \phi_{f(s)}$, $s \in S$. But this last equation simply asserts that $\phi j = f$.

The function j is evidently injective and we may use it to embed S in $\mathbf{Z}^{(S)}$; thus we may think of the free group on S as consisting of "words" in the elements of S.

Let G be a group and let S be a set of generators of G. There is then a homomorphism $\epsilon: F_S \rightarrow G$ which is the identity on S and it is plainly surjective. Thus *every group is the surjective image of a free group.* Let $F = F_S$ and let R be the kernel of $\epsilon: F \twoheadrightarrow G$. Then R is a (normal) subgroup of F and $F/R \cong G$. We now quote without proof a basic theorem of combinatorial group theory (see [8] for proofs).

THEOREM 6.3.   *A subgroup of a free group is free.*

Thus R is itself free. We call the pair (F, R) a (*free*) *presentation* of G, and S is then a set of generators of G. Now let T be a set of elements of R such that R is generated by the elements of T together with their conjugates (in F). We say that R is the *normal subgroup of F generated by* T, and we call T a set of *relators* for the set of generators S. Notice that the elements of T, that is, the relators, are words in the generators.

EXAMPLE 6.4.   Consider again the group P_3 of all permutations of three symbols $1, 2, 3$ (see Examples 3.5). From the group table it is easy to see that P_3 is generated by the set $S = (a_2, a_5)$. Write $a_2 = x, a_5 = y$. Then we may find a set of relators by writing down words in x, y which yield the neutral element of P_3 and are such that every other word yielding the neutral element is a consequence of the given words. It is not difficult to see that we may take for T the set of words $(x^2, y^3, (xy)^2)$.

Corresponding to each relator t there is the *relation* $t = 1$. However, if we specify relations instead of relators, we have an additional freedom, since we may express relations as equalities between words (in the generators), without insisting that one of the words be 1. Thus we may take as relations in x, y, for P_3,

$$x^2 = 1, \quad y^3 = 1, \quad xyx = y^2.$$

The following obvious property of free groups is often of crucial importance.

THEOREM 6.4. *Let $\eta : G \twoheadrightarrow H$ be surjective and let $\phi : F \to H$ be a homomorphism of the free group F into H. There is then a homomorphism $\psi : F \to G$ such that $\eta\psi = \phi$. [Diagram (6.1).]*

$$
\begin{array}{c}
F \\
\swarrow \quad \downarrow \phi \\
G \xrightarrow{\quad \eta \quad} H
\end{array}
\qquad (6.1)
$$

(Notice that here we do *not* claim uniqueness for ψ. For this reason, we do not write the symbol ψ into the diagram.)

PROOF. Let F be free on the set S. For each $s \in S$ choose $g \in G$ with $\eta(g) = \phi(s)$; this is possible since η is surjective. Set $f(s) = g$. By the universal property of free groups there is a unique homomorphism $\psi : F \to G$ such that $\psi | S = f$. But then $\eta\psi = \phi$ on S. It follows from the uniqueness of extensions from S to F that $\eta\psi = \phi$ throughout F.

In fact the converse of this theorem holds. For suppose that P is a group having the property claimed in (6.1) for F. Present P by $K \twoheadrightarrow P$, where K is free. As a special case of (6.1) for P, we infer $\mu : P \to K$ with $\epsilon\mu = 1$. But then μ is injective, and so μ embeds P as a subgroup of K. By Theorem 6.3, P is therefore itself free.

7. EXACT SEQUENCES

DEFINITION 7.1. We say that the sequence of groups and homomorphisms

$$
\cdots \to G_{n-1} \xrightarrow{f_{n-1}} G_n \xrightarrow{f_n} G_{n+1} \to \cdots , \quad -\infty < n < \infty
$$

is *exact at* G_n if ker $f_n = \mathrm{im} f_{n-1}$. It is *exact* if it is exact at G_n for all n. A *short exact sequence* (s.e.s) is an exact sequence

$$1 \to H \overset{\mu}{\to} G \overset{\epsilon}{\to} K \to 1. \tag{7.1}$$

Notice that to say that (7.1) is exact is to assert three things:

(i) μ is injective,

(ii) ϵ is surjective,

(iii) $\ker \epsilon = \operatorname{im} \mu$.

Thus μ maps H isomorphically onto the kernel of ϵ, so that μ is the kernel of ϵ, and ϵ is the cokernel of μ. Thus (7.1) is an abstract expression of the situation of group G, normal subgroup H, and factor group K. An interesting example of (7.1) is furnished by a presentation of K.

An important special case of a s.e.s. is provided by the direct product of two groups $G \times H$. Let $\iota_2 : H \to G \times H$ be the embedding and let $p_1 : G \times H \to G$ be the projection.

THEOREM 7.2.   *The sequence* $1 \to H \overset{\iota_2}{\to} G \times H \overset{p_1}{\to} G \to 1$ *is exact.*

PROOF.   Certainly ι_2 is injective, and p_1 is surjective. Moreover

$$p_1(g,h) = e \Leftrightarrow g = e \Leftrightarrow (g,h) \in \operatorname{im} \iota_2,$$

so $\ker p_1 = \operatorname{im} \iota_2$.

However, it is certainly not the case in general that (7.1) implies $G \cong H \times K$. For example, the sequence

$$1 \to \mathbf{Z}_2 \overset{\mu}{\to} \mathbf{Z}_4 \overset{\epsilon}{\to} \mathbf{Z}_2 \to 1$$

is exact, where $\mu[1] = [2], \epsilon[1] = [1]$. However $\mathbf{Z}_4 \not\cong \mathbf{Z}_2 \times \mathbf{Z}_2$, since $\mathbf{Z}_2 \times \mathbf{Z}_2$ has no element of order 4. It is a difficult and important problem to classify, for given H, K, the groups G which fit into a s.e.s. (7.1). In this section we are content to study conditions under which (7.1) is effectively obtained from the expression of G as the direct product of H and K, and to make a generalization.

We should first say what we understand by "effectively" in the preceding sentence. Suppose that we have the commutative diagram (7.2)

$$
\begin{array}{ccccccccc}
1 & \to & H & \overset{\mu}{\to} & G & \overset{\epsilon}{\to} & K & \to & 1 \\
  &   & \| & & \downarrow{\scriptstyle \phi} & & \| & & \\
1 & \to & H & \underset{\mu'}{\to} & G' & \underset{\epsilon'}{\to} & K & \to & 1
\end{array}
\tag{7.2}
$$

(that is, $\phi\mu = \mu', \epsilon = \epsilon'\phi$) in which ϕ is an isomorphism. It would then plainly be absurd to distinguish between the two s.e.s., and we would regard them as equivalent. We are thus going to be interested in the question whether, given (7.1), there exists (7.2) in which the bottom row is just $1 \to H \overset{\iota_1}{\to} H \times K \overset{p_2}{\to} K \to 1$. We will then say that (7.1) is *equivalent to a direct product*.

We can make an immediate simplification of any test we devise by means of the following observation.

PROPOSITION 7.3. *Given the diagram* (7.2), ϕ *is an isomorphism.*

In other words, it is superfluous to add the requirement that ϕ be an isomorphism.

PROOF. We first prove ϕ surjective. Let $g' \in G'$. Then, since ϵ is surjective,

$$\epsilon'g' = \epsilon g, \qquad \text{for some } g \in G,$$

$$= \epsilon'\phi g.$$

But $\ker \epsilon' = \operatorname{im} \mu'$, so that

$$g' = \phi g \cdot \mu' h, \qquad \text{for some } h \in H,$$

$$= \phi g \cdot \phi \mu h$$

$$= \phi(g \cdot \mu h).$$

Next we prove ϕ injective. Let $\phi g = e, g \in G$. Then $\epsilon g = \epsilon'\phi g = e$, so that, since $\ker \epsilon = \operatorname{im} \mu, g = \mu h, h \in H$. Then $e = \phi g = \phi \mu h = \mu' h$, so that $h = e$, since μ' is injective. Thus $g = \mu e = e$, and the theorem is proved.

THEOREM 7.4. *The s.e.s.* (7.1) *is equivalent to a direct product if and only if there exists* $\eta : G \to H$ *with* $\eta\mu = 1$.

PROOF. Clearly the s.e.s. $1 \to H \overset{\iota_1}{\to} H \times K \overset{p_2}{\to} K \to 1$ admits $p_1 : H \times K \to H$ with $p_1\iota_1 = 1$, so that the condition is necessary. Conversely, suppose the condition satisfied and consider the diagram

$$
\begin{array}{ccccccccc}
1 & \to & H & \overset{\mu}{\to} & G & \overset{\epsilon}{\to} & K & \to & 1 \\
& & \| & & \downarrow{\scriptstyle\{\eta,\epsilon\}} & & \| & & \\
1 & \to & H & \overset{\iota_1}{\to} & H \times K & \overset{p_2}{\to} & K & \to & 1
\end{array}
$$

Then $\{\eta, \epsilon\}\mu = \{\eta\mu, \epsilon\mu\} = \{1, 0\} = \iota_1$ and $p_2\{\eta, \epsilon\} = \epsilon$.

In view of Proposition 7.3, this establishes the theorem. We call η a *left-splitting* of (7.1) so that we may say that (7.1) is left-split if and only if it is equivalent to a direct product.

Plainly there is also the notion of a *right-splitting* of (7.1), that is, a homomorphism $\nu : K \to G$ such that $\epsilon\nu = 1$. Since the direct product sequence plainly has a right-splitting $\iota_2 : K \to H \times K$, it follows that the existence of a left-splitting implies the existence of a right-splitting. However, the converse is false. For consider the group P_3 (see Section 3) and its normal subgroup $\mathbf{Z}_3$ generated by the permutation a_4. We thus have the s.e.s.

$$1 \to \mathbf{Z}_3 \xrightarrow{\mu} P_3 \xrightarrow{\epsilon} \mathbf{Z}_2 \to 1, \tag{7.3}$$

where μ is the embedding and, if $\mathbf{Z}_2$ is generated by a, ϵ is given by

$$\epsilon a_1 = e, \qquad (a_1 = e \text{ in } P_3)$$

$$\epsilon a_2 = a,$$

$$\epsilon a_3 = a,$$

$$\epsilon a_4 = e,$$

$$\epsilon a_5 = e,$$

$$\epsilon a_6 = a.$$

Then a right-splitting of (7.3) is given by $\nu a = a_2$; but P_3 is certainly not isomorphic to $\mathbf{Z}_3 \times \mathbf{Z}_2$.

If a right-splitting $\nu : K \to G$ of (7.1) is given, it enables us to define an *action* of K on H, that is, a homomorphism $\omega : K \to \operatorname{Aut} H$, where $\operatorname{Aut} H$ is the group of automorphisms of H (see Exercise 4.9). Thus we define

$$\omega(k)(h) = \nu(k)h\nu(k)^{-1}. \tag{7.4}$$

It is easy to verify that ω, so defined, is a homomorphism $K \to \operatorname{Aut} H$; we will write koh for $\omega(k)(h)$.

Now suppose, conversely, that an action ω of K on H is given. We define a group structure on the Cartesian product $H \times K$ by the rule

$$(h_1, k_1)(h_2, k_2) = (h_1(k_1 oh_2), k_1 k_2). \tag{7.5}$$

We leave it to the reader to verify that this is a group structure, noting only that the neutral element is (e, e) and that

$$(h, k)^{-1} = (k^{-1}oh^{-1}, k^{-1}).$$

We write the group $H \times_\omega K$ to stress its dependence on the action ω and call it the *semidirect product* of H and K (relative to ω). Notice that we have homomorphisms

$$\mu_1 : H \to H \times_\omega K, \qquad \text{given by } \mu_1(h) = (h, e),$$

$$\mu_2 : K \to H \times_\omega K, \qquad \text{given by } \mu_2(k) = (e, k),$$

$$\epsilon_2 : H \times_\omega K \to K, \qquad \text{given by } \epsilon_2(h, k) = k.$$

Moreover, μ_1 is injective and $\mu_1 H$ is normal in $H \times_\omega K$. Indeed, we have a s.e.s.

$$1 \to H \xrightarrow{\mu_1} H \times_\omega K \xrightarrow{\epsilon_2} K \to 1, \tag{7.6}$$

and μ_2 is a right-splitting of (7.6).

Now let (7.1) be a s.e.s. with right-splitting $\nu : K \to G$, determining an action ω by (7.4). We prove

THEOREM 7.5. *Under these circumstances, there exists* $\phi : H \times_\omega K \to G$ *such that the diagram*

$$
\begin{array}{ccccccccc}
1 & \to & H & \xrightarrow{\mu} & G & \underset{\nu}{\overset{\epsilon}{\rightleftarrows}} & K & \to & 1 \\
& & \| & & \uparrow{\scriptstyle\phi} & & \| & & \\
1 & \to & H & \xrightarrow{\mu_1} & H \times_\omega K & \underset{\mu_2}{\overset{\epsilon_2}{\rightleftarrows}} & K & \to & 1
\end{array}
$$

commutes.

PROOF. Set $\phi(h, k) = \mu h \cdot \nu k$. We must first show that ϕ is a homomorphism. Now

$$\phi((h_1, k_1)(h_2, k_2)) = \phi(h_1(k_1 o h_2), k_1 k_2)$$

$$= \mu h_1 \, \mu(k_1 o h_2) \nu(k_1 k_2)$$

$$= \mu h_1 \cdot \nu k_1 \cdot \mu h_2 \cdot \nu k_1^{-1} \cdot \nu(k_1 k_2), \quad \text{by (7.4)}$$

$$= (\mu h_1 \cdot \nu k_1)(\mu h_2 \cdot \nu k_2)$$

$$= \phi(h_1, k_1)\phi(h_2, k_2).$$

The relations $\phi\mu_1 = \mu$, $\phi\mu_2 = \nu$, $\epsilon\phi = \epsilon_2$ are now trivial. Thus the theorem is proved, so that a s.e.s. with a given right-splitting ν is equivalent to a semidirect product relative to the action ω induced by ν. We remark that the direct product of H and K may be regarded as the semidirect product relative to the trivial action ω (that is, $\omega = 0 : K \to \operatorname{Aut} H$).

A particularly interesting and important case of a s.e.s. (7.1) is that in which H is commutative; we call this a *s.e.s. with abelian kernel* or an *extension of K with abelian kernel*. In this case,

$$1 \to H \overset{\mu}{\to} G \overset{\epsilon}{\to} K \to 1, \qquad H \text{ commutative}, \tag{7.7}$$

we obtain an action of K on H without requiring a right-splitting. For any $k \in K$, let $k = \epsilon g$. We then set (viewing H as a normal subgroup of G)

$$koh = ghg^{-1}, \qquad h \in H. \tag{7.8}$$

To see that koh is well defined we must show it independent of the choice of g in $\epsilon^{-1}k$. If also $\epsilon g_1 = k$, then $g_1 = gh_1, h_1 \in H$, so that

$$g_1 hg_1^{-1} = gh_1 hh_1^{-1}g^{-1} = ghg^{-1},$$

since H is commutative. Thus (7.8) does yield an action of K on H. Of course, if (7.7) admits a right-splitting $\nu : K \to G$, then (7.8) coincides with the action induced by ν, since ν gives us a rule for picking, for each $k \in K$, an element g in $\epsilon^{-1}k$. It follows that when H is commutative and there *is* a right-splitting, the induced action is independent of the choice of splitting.

It is easy to construct an example of a s.e.s. without right-splitting. Thus we may take

$$1 \to \mathbf{Z} \overset{\mu}{\to} \mathbf{Z} \overset{\epsilon}{\to} \mathbf{Z}_2 \to 1 \tag{7.9}$$

where μ is multiplication by 2 if we think of $\mathbf{Z}$ as the (additive) group of integers. For the only homomorphism $\mathbf{Z}_2 \to \mathbf{Z}$ is the constant homomorphism, so that there can be no ν with $\epsilon\nu = 1$.

[Incidentally, when, as in (7.9), all groups in an exact sequence are commutative and are written additively, it is customary to write 0 for 1; thus:

$$0 \to \mathbf{Z} \to \mathbf{Z} \to \mathbf{Z}_2 \to 0.$$

This will be our convention in the next chapter.]

EXERCISES

1.1. Show that, if G is a group, the equations $ax = b, ya = b, a, b \in G$ have a *unique* solution in G.

1.2. Show that e is the only *idempotent* in G; that is, show that the equation $x^2 = x$ in G implies $x = e$. More generally, show that if, given b in G, there exists a in G such that $ab = a$, then $b = e$.

1.3. Let G be a group and $a \in G$. Prove that *either* the powers of a are all distinct *or* there exists a positive integer n such that $a^m = 1$ if and only if $n | m$. [In the former case we say that a is of *infinite order*, in the latter case that a is of *order n*.]

1.4. Show that if every nonneutral element of G is of order 2, then G is commutative.

1.5. Complete the proof of Proposition 1.2.

1.6. Show that all groups of order $\leqslant 5$ are commutative. Exhibit a commutative and a noncommutative group of order 6.

1.7. Let P_n be the group of all permutations on n symbols. What is the order of P_n? (We call P_n the *symmetric group* on n symbols.)

2.1. Show that a subgroup of a cyclic group is cyclic.

2.2. Describe explicitly the elements of the smallest subgroup of the group P_4 containing (2134), (1342) (see Example 3.5.3 for notation).

2.3. Let H, K be subgroups of G and let HK be the subset of G consisting of all elements hk with $h \in H, k \in K$. Show that HK is a subgroup if G is commutative. Is it a subgroup in general?

2.4. Describe all the subgroups of the group of symmetries of the square.

2.5. Let $0(n)$ be the group of orthogonal transformations of $\mathbf{R}^n$. Show how $0(n)$ may be regarded as a subgroup of $0(n+1)$.

2.6. Let P_n be the group of all permutations of n symbols. Show how P_n may be regarded as a subgroup of P_{n+1}.

3.1. Let H be a subgroup of G. Show that H is normal in G if and only if $a^{-1}ha \in H$ whenever $h \in H$, $a \in G$.

3.2. Show by an example that if K is normal in H and H normal in G, K may fail to be normal in G. Show that if H is normal in G and H is cyclic, then any subgroup of H is normal in G.

3.3. Let $H_1, H_2, \ldots, H_k$ be normal subgroups of G. Show that the smallest subgroup of G containing all of $H_1, H_2, \ldots, H_k$ is the set H consisting of "words" $h_1 h_2 \cdots h_k$, $h_i \in H_i$; and that H is normal in G.

3.4. Let H_1, H_2 be normal subgroups of G whose intersection consists of the neutral element alone. Show that $h_1 h_2 = h_2 h_1$, $h_1 \in H_1$, $h_2 \in H_2$.

3.5. Let $[G, G]$ be the subgroup of G generated by all *commutators* $g_1^{-1} g_2^{-1} g_1 g_2$, $g_1, g_2 \in G$. Show that if H is a subgroup of G containing $[G, G]$ then H is normal in G and G/H is commutative; and, conversely, that if H is a normal subgroup of G such that G/H is commutative, then $H \supseteq [G, G]$. ($[G, G]$ is called the *commutator subgroup* of G.)

3.6. Let H be normal in G and let Q be a subgroup of G/H. Show that there is a unique subgroup K of G, such that $H \subseteq K$ and $Q = K/H$, and that K is normal in G if and only if Q is normal in G/H.

3.7. What is the index of P_n in P_{n+1} (see Exercise 2.6)?

3.8. Show that if p is prime, then every group of order p is cyclic.

4.1. Show (see Proposition 4.4) that a homomorphism $\phi : G \to K$ is injective if, for some $k \in \phi G$, $\phi^{-1}k$ is a singleton.

4.2. Let H be a subset of G. Show that the subgroup structure on H (if it exists) is uniquely characterized as that group structure which renders the embedding $H \subseteq G$ a homomorphism. If H is normal in G, show that the quotient group structure on G/H is uniquely characterized as that group structure which renders the projection $G \twoheadrightarrow G/H$ a homomorphism.

4.3. Show that every homomorphism $\phi : G \to K$ may be factored as

$$ G \overset{\epsilon}{\twoheadrightarrow} L \overset{\mu}{\rightarrowtail} K, $$

with ϵ surjective, μ injective. Show that if

$$ G \overset{\epsilon_i}{\twoheadrightarrow} L_i \overset{\mu_i}{\rightarrowtail} K, \quad i = 1, 2, $$

are two such factorizations, then there exists a unique isomorphism $\theta : L_1 \cong L_2$ such that $\theta \epsilon_1 = \epsilon_2, \mu_2 \theta = \mu_1$.

4.4. Suppose that, for each pair of groups G, K a homomorphism $\gamma_{GK} : G \to K$ is picked such that

$$ \phi \gamma_{GK} = \gamma_{GL}, \qquad \text{for all } G, K, L, \ \phi : K \to L, $$

$$ \gamma_{GK} \psi = \gamma_{MK}, \qquad \text{for all } M, G, K, \ \psi : M \to G. $$

Show that γ_{GK} is the constant homomorphism.

4.5. Let $\phi, \psi : G \to K$ be homomorphisms and let $H \subseteq G$ consist of those $g \in G$ such that $\phi g = \psi g$. Show that H is a subgroup of G. Characterize $\iota : H \subseteq G$ in a way similar to the characterization of the kernel in Theorem 4.7. (We call H the *equalizer* of ϕ and ψ.)

4.6. (i) Let $\phi : G \to K$ be injective. Show that ϕ has the following property: given any M and homomorphisms $\alpha_1, \alpha_2 : M \to G$, then $\alpha_1 = \alpha_2$ if $\phi \alpha_1 = \phi \alpha_2$.

(ii) Let $\phi : G \to K$ be surjective. Show that ϕ has the following property: given any L and homomorphisms $\beta_1, \beta_2 : K \to L$, then $\beta_1 = \beta_2$ if $\beta_1 \phi = \beta_2 \phi$. (In fact the converses of these statements also hold, but we postpone the proof of that. Property (i) describes a *monomorphism* ϕ, or asserts that ϕ is *monic*; property (ii) describes an *epimorphism* ϕ, or asserts that ϕ is *epic*. Thus, in fact, in the *category of groups*,

$$\text{injective} \Leftrightarrow \text{monic},$$

$$\text{surjective} \Leftrightarrow \text{epic.})$$

4.7. Let H be a subgroup of G. Describe a set of generators of the normal hull $\hat{H}$ of H in G in terms of a set S of generators of H. Show that, if $\phi : K \to G$ is a homomorphism such that, given any M, and any $\alpha : G \to M$, then $\alpha\phi = 0$ only if $\alpha = 0$, we *cannot* deduce that ϕ is a surjection. What inference of this kind can we deduce?

4.8. Let H be a normal subgroup of G and K a subgroup of G. Show that HK (see Exercise 2.3) is a subgroup of G, that $H \cap K$ is normal in K, and that

$$K/H \cap K \cong HK/H.$$

4.9. Let G be a group. An *automorphism* of G is an isomorphism $G \cong G$. Show that the set $\text{Aut}\, G$ of automorphisms of G is a group under function composition. Let $a \in G$ and consider the function $g \mapsto aga^{-1}$ from G to G. Show that this function ϕ_a is an automorphism of G and that $a \mapsto \phi_a$ is a homomorphism $\tau : G \to \text{Aut}\, G$. We call τG the group $\text{Inn}\, G$ of *inner* automorphisms of G. Identify the kernel of τ as the center of G. Show that $\text{Inn}\, G$ is normal in $\text{Aut}\, G$.

4.10. Show that a cyclic group is isomorphic either to $\mathbf{Z}$ or to $\mathbf{Z}_n$ for a given n. (Thus it is *infinite cyclic* or *cyclic of order n*, for some n.)

4.11. Enumerate, up to isomorphism, all groups of order $4, 6, 8$.

5.1. Show that $\mathbf{Z}_m \times \mathbf{Z}_n \cong \mathbf{Z}_{mn}$ if and only if m, n are mutually prime.

5.2. Show that ΠG_i is commutative if and only if each G_i is commutative. Show that the commutator subgroup (see Exercise 3.5) of ΠG_i is $\Pi[G_i, G_i]$, and that the center of ΠG_i is ΠC_i, where C_i is the center of G_i.

5.3. Show that if H_i is normal in G_i, then ΠH_i is normal in ΠG_i and

$$\Pi G_i / \Pi H_i \cong \Pi(G_i/H_i).$$

5.4. Show that the free product of (more than one) nontrivial groups is never commutative.

5.5. Give an explicit description of $\omega : \amalg\, G_i \to \Pi G_i$. Show that the kernel of $\omega : G_1 \amalg G_2 \to G_1 \times G_2$ is the subgroup of $G_1 \amalg G_2$ generated by commutators $gg'g^{-1}g'^{-1}$, where $g \in q_1(G_1), g' \in q_2(G_2)$.

6.1. Let S be a subset of the group G and let $\phi: F_S \to G$ be the homomorphism attested by the universal property of the free group on S. Describe in familiar terms the properties of S corresponding to the statements (i) ϕ is surjective, (ii) ϕ is injective, and (iii) ϕ is bijective.

6.2. Let the groups G_i be given by generators and relators. Describe $\amalg\ G_i$ by generators and relators.

6.3. Show that the kernel of $\omega: G_1 \amalg G_2 \to G_1 \times G_2$ is free on the commutators $gg'g^{-1}g'^{-1}$, where $e \neq g \in q_1(G_1), e \neq g' \in q_2(G_2)$. (See Exercise 5.5.)

6.4. Give a set of generators and relators for (i)$\mathbf{Z}_n$, (ii)P_4, and (iii) the symmetry group of the square.

6.5. Show that a free group has no nonneutral elements of finite order. Show that the group $\mathbf{Q}$ of rationals is not free.

6.6. Let S be a set of generators of the group G and let T be a set of relators (words in the elements of S). Let $f: S \to H$ be a function from S to the underlying set of the group H. Show that f extends to a unique homomorphism $\phi: G \to H$ if, and only if, the extension $\bar{f}$ of f to a homomorphism $F_S \to H$ maps every relator to 1_H.

7.1. Show that $\cdots \to G_{n-1} \xrightarrow{\phi_{n-1}} G_n \xrightarrow{\phi_n} G_{n+1} \to \cdots$ is exact at G_n if and only if the induced sequence

$$1 \to \phi_{n-1}G_{n-1} \to G_n \to \phi_n G_n \to 1$$

is a s.e.s.

7.2. Construct an explicit inverse to ϕ in Theorem 7.5.

7.3. Let $1 \to H \to G \to K \to 1$ be a right-split s.e.s. with H in the center of G. Show that the sequence is equivalent to the direct product.

7.4. Call a permutation in P_n *even* if it is a product of an even number of *transpositions* (ij), $1 \leqslant i < j \leqslant n$. Let A_n be the subgroup consisting of even permutations (it is called the *alternating* group on n symbols). Show that A_n is of index 2 in P_n, and that P_n is the semidirect product of A_n and $\mathbf{Z}_2$.

7.5. Show how to associate with the s.e.s. $1 \to H \to G \to K \to 1$ a homomorphism $\tilde{\omega}: K \to \operatorname{Aut} H / \operatorname{Inn} H$ (see Exercise 4.9).

7.6. Let K act on H. A *derivation* $d: K \to H$ is a function such that

$$d(k_1 k_2) = dk_1(k_1 o dk_2), \quad k_1, k_2 \in K.$$

If $\phi: X \to K$ is a homomorphism then X acts on H through ϕ by $xoh = \phi(x)oh$. A ϕ-*derivation* $X \to H$ is then a derivation for this action. Show that $p_1: H \times_\omega K \to H$ is a p_2-derivation. Show that any homomorphism $\psi: X \to H \times_\omega K$ yields a homomorphism

$p_2\psi : X \to K$ and a $p_2\psi$-derivation $p_1\psi : X \to H$. Show, conversely, that, given a homomorphism $\phi : X \to K$ and a ϕ-derivation $d : X \to H$, there is a unique homomorphism $\psi : X \to H \times_\omega K$ such that $\phi = p_2\psi, d = p_1\psi$. Hence characterize $H \times_\omega K$ by a universal property. Set up a one-one correspondence between the set of derivations $K \to H$ and a certain subset of the set of homomorphisms $K \to H \times_\omega K$.

2

ABELIAN GROUPS

In this chapter our main concern is to demonstrate certain important respects in which abelian group theory differs from group theory. We do not even take the theory of abelian groups as far as we plan to go in this book, as we prefer to postpone much of what we wish to say to Chapters 4 and 5, where we have available the language of categories and functors and can subsume much of the theory under our more general study of modules. Thus the present chapter looks back to group theory and forward to module theory; and it provides useful examples of many of the ideas introduced formally in Chapter 3.

1. SPECIAL FEATURES OF COMMUTATIVE GROUPS

In this section we look at the development of (elementary) group theory that we discussed in Sections 1 to 4 of Chapter 1, and draw attention to places where the fact that in this chapter we are concerned exclusively with commutative groups essentially simplifies—or in some other way changes —the situation.

The first, and very crucial, simplification is that *every subgroup of a commutative group is normal*. Thus given any commutative group G and subgroup H we may form the factor group G/H. Of course, if G is commutative, so are H and G/H so that these concepts are proper to the theory of commutative groups.

When we turn to the consideration of homomorphisms, many new features present themselves. Let G, K be commutative groups and let $\text{Hom}(G, K)$ be the set of homomorphisms from G to K. Then we may endow $\text{Hom}(G, K)$ with a commutative group structure by *elementwise multiplication*; that is, we set

$$(\phi \cdot \psi)(g) = \phi(g)\psi(g), \quad \phi, \psi \in \text{Hom}(G, K), \quad g \in G. \quad (1.1)$$

Let us check that $\phi \cdot \psi$ is indeed a homomorphism; we have

$$(\phi \cdot \psi)(gg') = \phi(gg')\psi(gg')$$

$$= \phi(g)\phi(g')\psi(g)\psi(g'), \quad \text{since } \phi, \psi \text{ are homomorphisms,}$$

$$= \phi(g)\psi(g)\phi(g')\psi(g'), \quad \text{since } K \text{ is commutative}$$

$$= (\phi \cdot \psi)(g)(\phi \cdot \psi)(g').$$

Moreover, the commutativity of K again makes it obvious from (1.1) that $\phi \cdot \psi = \psi \cdot \phi$. The other group axioms are readily verified; the product (1.1) is associative because K is associative; the constant homomorphism is the neutral element in $\text{Hom}(G,K)$; and ϕ^{-1} is given by

$$\phi^{-1}(g) = \phi(g)^{-1}. \tag{1.2}$$

Again, (1.2) defines a *homomorphism* ϕ^{-1} because K is commutative.

Note that no use has been made in the argument above of the fact that G is commutative. Thus $\text{Hom}(G,K)$ is a commutative group if G is a group and K a commutative. This generalization is not, however, very significant, for it is easy to see that, if K is commutative, every homomorphism $\phi: G \rightarrow K$ induces a homomorphism $\bar{\phi}: G/[G,G] \rightarrow K$ by the rule

$$\bar{\phi}(g[G,G]) = \phi g,$$

and $\phi \mapsto \bar{\phi}$ sets up an isomorphism

$$\text{Hom}(G,K) \cong \text{Hom}(G/[G,G],K). \tag{1.3}$$

Thus the study of homomorphisms from groups to commutative groups is reduced, by (1.3), to the study of homomorphisms from commutative groups to commutative groups. (Of course, $G/[G,G]$ is commutative; see Exercise 3.5 of Chapter 1.)

We emphasize that if K is not commutative then $\text{Hom}(G,K)$ does not even acquire a *group* structure through (1.1); for, in general, $\phi \cdot \psi$ is *not* a homomorphism unless K is commutative. Of course, for a particular choice of ϕ, ψ, it may turn out that $\phi \cdot \psi$ is a homomorphism, but it is not difficult to show that if for *all* groups G and homomorphisms $\phi, \psi: G \rightarrow K$, the function $\phi \cdot \psi$ defined by (1.1) is a homomorphism, then K must be commutative.

It might be thought that conceivably there may be some other reasonable rule whereby $\text{Hom}(G,K)$ may be given a group structure, when G and K are groups. This is not so, if we use a definition of "reasonable" suggested by the following proposition.

PROPOSITION 1.1.   *Let* G, G', K, K' *be commutative groups. Then, for all* $\phi, \psi : G \to K$,

$$(\phi \cdot \psi) o \alpha = (\phi o \alpha) \cdot (\psi o \alpha), \qquad \text{for all } \alpha : G' \to G,$$

$$\beta o (\phi \cdot \psi) = (\beta o \phi) \cdot (\beta o \psi), \qquad \text{for all } \beta : K \to K'.$$

where o indicates function composition.

PROOF

$$(\phi \cdot \psi o \alpha)(x) = (\phi \cdot \psi)(\alpha x) = \phi \alpha (x) \psi \alpha (x), \qquad \text{by (1.1)},$$

$$= (\phi \alpha \cdot \psi \alpha)(x), \qquad x \in G'.$$

On the other hand,

$$(\beta o (\phi \cdot \psi))(y) = \beta (\phi (y) \psi (y)), \qquad \text{by (1.1)},$$

$$= \beta \phi (y) \beta \psi (y), \qquad \text{since } \beta \text{ is a homomorphism,}$$

$$= (\beta \phi \cdot \beta \psi)(y), \qquad y \in G.$$

Notice that, in *this* argument, no use is made of commutativity. We show in the next chapter that no choice of group structure in the sets $\text{Hom}(G, K)$, as G, K range over the category of groups, is possible that satisfies the *distributive laws* expressed in Proposition 1.1.

We may state Proposition 1.1 in different form. The homomorphism $\alpha : G' \to G$ induces a function $\alpha^* : \text{Hom}(G, K) \to \text{Hom}(G', K)$ by the rule

$$\alpha^* (\phi) = \phi \alpha \tag{1.4}$$

[see Proposition 4.2(ii) of Chapter 1]. Similarly, the homomorphism $\beta : K \to K'$ induces a function $\beta_* : \text{Hom}(G, K) \to \text{Hom}(G, K')$ by the rule

$$\beta_* (\phi) = \beta \phi. \tag{1.5}$$

The we may restate Proposition 1.1 as

PROPOSITION 1.2.   *Let* G, G', K, K' *be commutative groups,* $\alpha : G' \to G$, $\beta : K \to K'$ *homomorphisms. Then* $\alpha^* : Hom(G, K) \to Hom(G', K)$ *and* $\beta_* : Hom(G, K) \to Hom(G, K')$ *are homomorphisms.*

Finally, we note that if G is commutative, then diagram (4.5) of Chapter 1 is valid for any homomorphism $\phi : K \to G$ and provides the universal defining property of the *cokernel* $\pi : G \to G/\phi K$ of ϕ. Thus we have, conceptually, no additional complication in describing the cokernel, over the description of the kernel. We also note, without proof, the following neat facts.

PROPOSITION 1.3. *In the category of commutative groups, every injective homomorphism is the kernel of its cokernel, and every surjective homomorphism is the cokernel of its kernel.*

Recall that, for groups, the first statement is false (since not all subgroups are normal), while the second statement is true.

We defined in Exercise 4.6 of Chapter 1 the important properties of *monomorphism* and *epimorphism*. We insist, from now on, that all groups entering our discussion are *commutative*, and prove

THEOREM 1.4. *In the category of commutative groups, the monomorphisms are precisely the injective homomorphisms and the epimorphisms are precisely the surjective homomorphisms.*

PROOF. We have already noted that injective homomorphisms are monomorphisms, so we must prove the converse. Let $\phi: G \to K$ be a monomorphism and let $\phi(g) = \phi(g')$. Let $\alpha, \beta: \mathbf{Z} \to G$ be the homomorphisms given by $\alpha(1) = g, \beta(1) = g'$. Then $\phi\alpha = \phi\beta$. Since ϕ is a monomorphism, $\alpha = \beta$, so that $g = g'$, and ϕ is injective.

We have already noted that surjective homomorphisms are epimorphisms, so we must prove the converse. Let $\phi: G \to K$ be an epimorphism and let $\pi: K \to K/\phi G$ be its cokernel (note that our groups are commutative). Then $\pi\phi = 0\phi = 0: G \to K/\phi G$. Since ϕ is an epimorphism, $\pi = 0$, so that $K = \phi G$ and ϕ is surjective.

We remark that the argument equating monomorphisms with injective homomorphisms is valid in the category of groups, while the argument equating epimorphisms with surjective homomorphisms is not. Indeed, it is a subtler fact that epimorphisms coincide with surjective homomorphisms in the category of groups.

The following fact (or, rather, half of it) sharply distinguishes between the case of groups and that of commutative groups.

THEOREM 1.5. *Let* $\phi: G \to K$ *be a homomorphism of commutative groups. Then*
  (i) *ϕ is injective if and only if $\phi\alpha = 0 \Rightarrow \alpha = 0$,*
  (ii) *ϕ is surjective if and only if $\beta\phi = 0 \Rightarrow \beta = 0$.*

PROOF. Let us write all groups additively; note that this has the desirable notational effect that the constant homomorphism from G to K, which we have been denoting by $0: G \to K$, is the neutral element (or "zero element") of the *additive* abelian group $\mathrm{Hom}(G, K)$.

If ϕ is a monomorphism and $\phi\alpha = 0$, then $\alpha = 0$. If $\phi\alpha = 0 \Rightarrow \alpha = 0$ and $\phi\alpha_1 = \phi\alpha_2$ then (Proposition 1.2) $\phi(\alpha_1 - \alpha_2) = 0$, so that $\alpha_1 - \alpha_2 = 0$, $\alpha_1 = \alpha_2$, and ϕ is a monomorphism. Similarly we prove that $\beta\phi = 0 \Rightarrow \beta = 0$ is equivalent to ϕ being an epimorphism, so that we may apply Theorem 1.4.

As hinted, if $\phi: G \to K$ is a homomorphism of groups, then (i) holds, but (ii) is false (see Exercise 4.7 of Chapter 1).

Henceforward, we will always write our groups *additively*, so that we may properly speak of *abelian groups*.

2. DIRECT SUMS AND PRODUCTS OF ABELIAN GROUPS; THE STRUCTURE OF FINITELY GENERATED ABELIAN GROUPS

If we turn to Section 5 of Chapter 1 we remark that the discussion there of the direct product of groups applies in its entirety, but with significant simplification, to the case of abelian groups. For the direct product of commutative groups is again commutative, and Theorem 5.1 of that chapter holds in the context of abelian groups. Theorem 5.2, however, is much simplified because we do not have to worry about normality. Moreover, we may also form, as there, the restricted direct product and thus obtain a generalization of the theorem. Thus we will feel free to quote the results of Section 5 of Chapter 1, insofar as they relate to the direct product.

However, when we come to consider the free product, a very different situation faces us. For the free product of (two or more) nontrivial groups is never commutative. Thus we cannot, as in the case of the direct product, merely confine attention to commutative groups in order to obtain a satisfactory theory of free products of abelian groups. We may certainly formulate property F, as in Section 5 of Chapter 1, but now only with respect to homomorphisms of abelian groups; and we now must seek a new existence proof of a suitable collection $(\tilde{A}; q_i)$ where A_i is a family of abelian groups and $q_i: A_i \to \tilde{A}$ are homomorphisms. That is, we must modify the construction given in Chapter 1 in order that $\tilde{A}$ be commutative.

In fact, the situation turns out to be much simpler for abelian groups than for groups. Let us write $\oplus A_i$ for the *restricted* direct product of the abelian groups A_i and let us write $q_i: A_i \to \oplus A_i$ for the homomorphisms ι_i of Section 5 of Chapter 1 with the range reduced from ΠA_i to $\oplus A_i$.

THEOREM 2.1. *The collection $(\oplus A_i; q_i)$ has property F with respect to homomorphisms to abelian groups.*

PROOF. The theorem asserts that if $\xi_i: A_i \to X$ is a family of homomorphisms from A_i to the abelian group X, then there exists a unique homomorphism $\xi: \oplus A_i \to X$ with $\xi q_i = \xi_i$. Now, since $\oplus A_i$ is the smallest subgroup of ΠA_i containing all $q_i A_i$, it is plain that ξ, if it exists, is unique. Indeed we are forced by the relations $\xi q_i = \xi_i$ to define ξ by

$$\xi\{a_i\} = \sum \xi_i a_i. \tag{2.1}$$

Now (2.1) makes sense, since the sum on the right is finite ($a_i = 0$ except for a finite number of indices i). Moreover (2.1) *does* define a homomorphism, since X is commutative. Thus the theorem is proved.

REMARK. The reader may have expected us to construct the "free product in the category of abelian groups" by taking the free product in the category of groups, ΠA_i, and factor out the commutator subgroup $[\Pi A_i, \Pi A_i]$ [see (1.3)]. In fact, this does lead to the restricted direct product, so that the procedure would, indeed, have been valid. The frontal attack is, however, simpler, since it means that, if our discussion is to be concentrated on the category of abelian groups, there is no need to mention the—somewhat complicated—free product at all.

DEFINITION 2.1.   The restricted direct product of abelian groups is called the *direct sum*. If $A = \oplus_i A_i$, then each A_i is called a *direct summand* in A.

Notice, in particular, that if the indexing set is finite, say $I = (1, 2, \ldots, k)$, then

$$\Pi A_i = \oplus A_i,$$

and we would write both as

$$A_1 \oplus A_2 \oplus \cdots \oplus A_k.$$

Of course, strictly speaking, we should not say that, even then, the direct sum and direct product coincide, since we regard the direct sum as $(\oplus A_i; q_i)$ and the direct product as $(\oplus A_i; p_i)$.

For abelian groups, we have a particularly satisfactory structure theorem in the finitely generated case, which we state in this section and prove in Chapter 5. We first need the notion of a *free abelian* group. This is defined just as a free group was defined in Section 6 of Chapter 1, except that we now demand the universal property for functions to *abelian* groups G. It is then easy to see that the free abelian group on the set S is just the direct sum of copies of $\mathbf{Z}$ indexed by S. We write this $\mathbf{Z}^{[S]}$ and, as in Chapter 1, we may think of S as a basis for $\mathbf{Z}^{[S]}$ which, in particular, generates $\mathbf{Z}^{[S]}$. Obviously $\mathbf{Z}^{[S]}$ and $\mathbf{Z}^{[S']}$ are isomorphic if and only if S and S' have the same cardinal; this cardinal is called the *rank* of $\mathbf{Z}^{[S]}$. We describe S, or the image of S in $\mathbf{Z}^{[S]}$, as a *basis* for $\mathbf{Z}^{[S]}$.

Now let A be an abelian group and consider the subset T of A consisting of elements of finite order. This is a subgroup; for if $m_1 a_1 = 0$, $m_2 a_2 = 0$, then $m_1 m_2 (a_1 - a_2) = m_1 m_2 a_1 - m_1 m_2 a_2 = 0$. (Notice that the corresponding statement for groups is false; consider, for example, $\mathbf{Z}_2 \amalg \mathbf{Z}_2$). Moreover the factor group A/T has no elements (except 0) of finite order. For if[†] $a + T$

[†]Since we are writing our groups additively, we write a coset as $a + T$ instead of aT.

is of finite order m, then $ma + T = T$, $ma \in T$, so ma is of finite order, whence a is of finite order, $a \in T$. We call T the *torsion subgroup* of A and say that A is a *torsion group* if $A = T$. An abelian group (like A/T) with no elements (except 0) of finite order is called *torsion-free*.

THEOREM 2.2. *If B is a finitely generated torsion-free abelian group then* B *is free abelian.*

This will be proved in Chapter 5 (Theorem 4.4).

COROLLARY 2.3. *If A is an abelian group with torsion subgroup* T, *such that* A/T *is finitely generated, then*

$$A \cong T \oplus A/T.$$

This will be deduced from Theorem 3.3 in the next section.

Now suppose that A is finitely generated. Then certainly A/T is finitely generated, so that A/T is free abelian by Theorem 2.2, and we define $\rho(A)$, the rank of A, by

$$\rho(A) = \rho(A/T). \tag{2.2}$$

Moreover it follows from Corollary 2.3 that T is finitely generated. Thus as a *finitely generated abelian group, all of whose elements are of finite order*, T must be *finite*! Corollary 2.3 has therefore reduced the study of the structure of finitely generated abelian groups to that of the study of finite abelian groups.

Now let p be a prime, and let $T^{(p)}$ be the subset of T consisting of those elements whose orders are a power of p. It is then easy (as above) to prove that $T^{(p)}$ is a subgroup of T. If $T^{(p)} \neq 0$, we say that T has p-torsion; if T is the torsion subgroup of A, we also say that A has p-torsion if $T^{(p)} \neq 0$.

THEOREM 2.4. T *is the* (*internal*) *direct sum of its subgroups* $T^{(p)}$.

PROOF. Applying Theorem 5.2 of Chapter 1 in the abelian case, we simply have to show that every element x in T is uniquely expressible as

$$x = \sum_p x_p, \quad x_p \in T^{(p)}. \tag{2.3}$$

Of course, this sum would in any case be finite but, since T is a finite group, only a finite number of primes in fact enter into the discussion.

Now let x be of order n and let n have the prime power decomposition

$$n = p_1^{m_1} \cdots p_k^{m_k}.$$

Let $q_i = n/p_i^{m_i}, i = 1, 2, \ldots, k$. Then $q_i x$ is of order $p_i^{m_i}$, so that $q_i x \in T^{(p_i)}$. Moreover,

$$gcd(q_1, q_2, \ldots, q_k) = 1,$$

so that we may find integers $m_1, m_2, \ldots, m_k$, with

$$1 = m_1 q_1 + m_2 q_2 + \cdots + m_k q_k.$$

Thus

$$x = m_1 q_1 x + m_2 q_2 x + \cdots + m_k q_k x,$$

with

$$m_i q_i x \in T^{(p_i)},$$

so that every x in T admits an expression (2.3).

To prove uniqueness, suppose that $0 = x_1 + x_2 + \cdots + x_k$, $x_i \in T^{(p_i)}$, where $p_1, p_2, \ldots, p_k$ are distinct primes; we want to show that $x_1 = x_2 = \cdots = x_k = 0$. Let the order of x_i be $p_i^{n_i}$. Now

$$x_1 = -x_2 - x_3 - \cdots - x_k$$

and $p_1^{n_1} x_1 = 0$, while $p_2^{n_2} \ldots p_k^{n_k}(-x_2 - x_3 - \cdots - x_k) = 0$. Since $gcd(p_1^{n_1}, p_2^{n_2} \ldots p_k^{n_k}) = 1$, it follows that $x_1 = 0$. Similarly $x_2 = \cdots = x_k = 0$, and the theorem is proved.

Theorem 2.4 further reduces the study of the structure of finitely generated abelian groups to that of finite abelian *p-groups*, a *p*-group being a group all of whose elements are of order a power of a fixed prime *p*.

We will prove in Chapter 5

THEOREM 2.5. *Let* $T^{(p)}$ *be a finite abelian p-group. Then*

$$T^{(p)} \cong \mathbf{Z}_{p^{n_1}} \oplus \mathbf{Z}_{p^{n_2}} \oplus \cdots \oplus \mathbf{Z}_{p^{n_k}}, \tag{2.4}$$

where $n_1 \geqslant n_2 \geqslant \cdots \geqslant n_k$, *and* $n_1, n_2, \ldots, n_k$ *are entirely determined by* $T^{(p)}$.

Summing up, we have the

Structure Theorem for Finitely Generated Abelian Groups. *A finitely generated abelian group* A *is the direct sum of a free abelian group of rank* $\rho(A)$ *and finite abelian p-groups* $T^{(p)}$, *expressible as* (2.4). *The isomorphism class of* A *is determined by* $\rho(A)$ *together with the positive integers* $n_1(p), n_2(p), \ldots,$ $n_{k(p)}(p)$, *where* p *ranges over the (finite) set of primes such that* A *has p-torsion.*

Applications of this theorem are given in the exercises.

We close this section with a discussion of free (abelian) *presentations* of abelian groups; this also is modeled on the discussion in Section 6 of Chapter 1. If S is a set of generators of G, we have a unique surjection $\epsilon : Z^{[S]} \twoheadrightarrow G$ which is the identity on S. The kernel of ϵ is a subgroup of $Z^{[S]}$ and, therefore, as we prove in the next section, a free abelian group. Thus every abelian group G may be presented as the quotient group of a free abelian group F by a subgroup R (which is also free abelian), $G \cong F/R$. If R is free abelian on the set T, we call T a set of *relators* (in the generators S) for the abelian group G. Even if T merely generated R, we would say that T is a set of relators for G; by taking T to be a basis for R, we assure ourselves a *nonredundant* set of relators, or *minimal* set of relators.

If S is finite then one may prove that a minimal relator set T is finite (see Theorem 3.6); that is, an abelian group with a finite set of generators has a finite set of relators. This is in marked contrast to the situation for groups, where there are finitely generated groups that have no presentation with a finite set of relators.

3. PROJECTIVE AND INJECTIVE ABELIAN GROUPS

We begin this section by noting the following analog of Theorem 6.4 of Chapter 1, whose proof we leave to the reader.

THEOREM 3.1. *Let* $\eta : A \twoheadrightarrow B$ *be a surjective homomorphism of abelian groups and* $\phi : F \to B$ *a homomorphism from the free abelian group* F *to* B. *Then there is a homomorphism* $\psi : F \to A$ *with* $\eta\psi = \phi$.

Let us say that an abelian group having the property claimed for F in this theorem is *projective*, so that the theorem claims that free abelian groups are projective. We show later in this section that the converse holds. Nevertheless we first deduce some elementary facts about projective abelian groups since they hold in the more general context of Λ-modules discussed in Chapter 4, where projective modules are not necessarily free.

PROPOSITION 3.2.
   (*i*) *A direct sum of projective abelian groups is projective.*
   (*ii*) *A direct summand in a projective abelian group is projective.*

PROOF.

(i) Let $P = \oplus P_i$, each P_i projective and consider $\eta : A \twoheadrightarrow B, \phi : P \to B$. Then $\phi = \langle \phi_i \rangle$, $\phi_i : P_i \to B$ and, P_i being projective, we find $\psi_i : P_i \to A$ with $\eta \psi_i = \phi_i$. Then $\eta \psi = \phi$, where $\psi = \langle \psi_i \rangle$.

(ii) Let P be projective and $P = Q \oplus R$; we prove Q projective. We have $q : Q \to P, p : P \to Q$ with $pq = 1$. Given $\phi : Q \to B, \eta : A \twoheadrightarrow B$, set $\bar{\phi} = \langle \phi, 0 \rangle = \phi p : P \to B$. Since P is projective, we find $\bar{\psi} : P \to A$ with $\eta \bar{\psi} = \bar{\phi}$. Then, if $\psi = \bar{\psi} q : Q \to A$, we have $\eta \psi = \eta \bar{\psi} q = \bar{\phi} q = \phi p q = \phi$, so Q is projective.

THEOREM 3.3. *The following statements are equivalent:*

(i) P is a projective abelian group.

(ii) P is a direct summand in a free abelian group.

(iii) To every surjection $\eta : A \twoheadrightarrow P$, there exists $\nu : P \rightarrowtail A$ with $\eta \nu = 1$.

(iv) To every surjection $\eta : A \twoheadrightarrow P$, there exists $\nu : P \rightarrowtail A$ with $A = \nu P \oplus \ker \eta$.

PROOF.

(i)$\Rightarrow$(iii) Trivial from the definition of projectivity.

(iii)$\Rightarrow$(iv) A special case of Theorem 4.1 below.

(iv)$\Rightarrow$(ii) We present P by $\epsilon : F \twoheadrightarrow P$ with F free abelian. Then P is embedded as a direct summand in F by ν.

(ii)$\Rightarrow$(i) F, being free, is projective by Theorem 3.1. Then P, a direct summand in F, is projective by Proposition 3.2(ii).

We note now that Corollary 2.3 follows from Theorem 3.3, since A / T, being free abelian, is projective; we apply (iv).

The assertions thus far in this section easily generalize to modules over any (unitary) ring; they reappear in this generalized form in Chapter 4. However, our next result does not generalize to modules over any ring, but, as we see later, requires severe restrictions on the ring. The reader should also note the analogy with Theorem 6.3 of Chapter 1.

THEOREM 3.4. *A subgroup of a free abelian group is free abelian.*

PROOF. We prove this theorem by transfinite induction. The reader who is unfamiliar with this aspect of set theory may be content to assume that we are in the finitely generated case, when a simple induction suffices.

We suppose F to be free abelian on the set I, which we may assume to be well ordered. We write f_i for the basis element of F corresponding to $i \in I$. For each i in I let F_i be the subgroup of F generated by the f_j with $j \leqslant i$. Then $F_{i_1} \subseteq F_{i_2}$ if $i_1 \leqslant i_2$ and $F = \cup_i F_i$. Furthermore, let $R_i = R \cap F_i$, so that $R_{i_1} \subseteq R_{i_2}$ if $i_1 \leqslant i_2$ and $R = \cup_i R_i$. Let $\tau_i : F \to \mathbf{Z}$ be the function that assigns to $x \in F$ the coefficient of f_i in the unique expression for x as a linear combination of basis elements. Then $\tau_i(R_i)$ is a subgroup of $\mathbf{Z}$, so that

$$\tau_i(R_i) = m_i \mathbf{Z}, \qquad m_i \geqslant 0.$$

Let $J \subseteq I$ be the subset of I consisting of those i for which $m_i \neq 0$. If $i \in J$, let $r_i \in R_i$ be chosen with $\tau_i(r_i) = m_i$. We will show that the set $\{r_i, i \in J\}$ is a basis for R.

First, we prove that the r_i are linearly independent. If not, there is a relation

$$n_1 r_{i_1} + n_2 r_{i_2} + \cdots + n_k r_{i_k} = 0,$$

with $i_1 < i_2 < \cdots < i_k$ and $n_k \neq 0$. Apply τ_{i_k}; since, for $l < k, \tau_{i_k}(r_{i_l}) = 0$, we have

$$n_k m_{i_k} = 0,$$

which is a blatant contradiction.

Second, we prove that the r_i generate R. Since $R = \cup R_i$, it is plainly sufficient to show that, for any $i \in I$, the $r_j, j \leqslant i$, generate R_i, and we argue by induction. Let the $r_j, j \leqslant i$, generate R_i'. Since $r_j \in R_j \subseteq R_i$, $R_i' \subseteq R_i$, and we prove, by induction on i, that $R_i \subseteq R_i'$. There are actually two cases: (i) $i \notin J$, (ii) $i \in J$. Suppose that we are in case (i) and $y \in R_i$. Since $\tau_i(y) = 0$ it follows that, in fact, $y \in R_j$ for some $j < i$. Thus $y \in R_j \subseteq R_j'$ (by the inductive hypothesis) $\subseteq R_i'$, so that $R_i \subseteq R_i'$.

Now suppose that we are in case (ii) and $y \in R_i$. Since $\tau_i(R_i) = m_i \mathbf{Z}$, $\tau_i(y) = km_i$ for some integer k. Thus $\tau_i(y - kr_i) = 0$, so that $y - kr_i \in R_j$ for some $j < i$. Thus

$$y - kr_i \in R_j \subseteq R_j' \subseteq R_i',$$

but, of course, $r_i \in R_i'$, so that $y \in R_i', R_i \subseteq R_i'$, and the induction is complete.

COROLLARY 3.5. *An abelian group is free if and only if it is projective.*

For, in fact, by Theorem 3.3, every projective abelian group is a direct summand in a free abelian group.

If we examine the proof of Theorem 3.4 we see that the rank of R is the cardinality of J, so that, since $J \subseteq I, \rho(R) \leqslant \rho(F)$. Thus we may refine Theorem 3.4 to the statement:

THEOREM 3.6. *A subgroup* R *of a free abelian group* F *is free abelian with* $\rho(R) \leqslant \rho(F)$.

We now turn our attention to another class of abelian groups, for which no useful parallel exists in the category of groups. We wish to "dualize" (in a sense explained in Chapter 3) the concept of a projective abelian group. We are led to define an abelian group K as *injective* if, for any monomorphism of abelian groups $\mu: B \rightarrowtail A$ and homomorphism $\phi: B \to K$, there is a homomorphism $\psi: A \to K$ with $\psi\mu = \phi$,

(Notice the direction of the arrows, compared with the diagram of Theorem 3.1.) It is not yet clear that there are nontrivial abelian groups K with this property. Before proving that there are such injective groups, we enunciate "duals" of the elementary facts adduced earlier in this section for projective abelian groups. We attach the affix "d" to indicate how the pairing of statements takes place in this duality at which we are vaguely hinting.

PROPOSITION 3.2D.
   (i) *A direct product of injective abelian groups is injective.*
   (ii) *A direct factor of an injective abelian group is injective.*
We leave the proof to the reader.

THEOREM 3.3D.   *The following statements are equivalent:*
   (i) K *is an injective abelian group.*
   (ii) K *is a direct factor of a cofree abelian group.*
   (iii) *To every injection* $\nu : K \rightarrowtail A$, *there exists* $\eta : A \twoheadrightarrow K$ *with* $\eta\nu = 1$.
   (iv) *To every injection* $\nu : K \rightarrowtail A$, *there exists* $\eta : A \twoheadrightarrow K$ *with* $A = \nu K \oplus \ker \eta$.

If we study the proof of Theorem 3.3, we see that what we need to prove Theorem 3.3d are:
   (a) A definition of *cofree* abelian groups!
   (b) A proof that cofree abelian groups are injective.
   (c) A proof that every abelian group may be embedded in a cofree abelian group.
It is, however, important to see that, were we content to prove the equivalence of (i), (iii), and (iv) in Theorem 3.3d, it would be sufficient for us to prove

THEOREM 3.7.   *Every abelian group may be embedded in an injective abelian group.*
Our procedure is to prove Theorem 3.7 in the text, and to take the reader through steps (a), (b), and (c), above, in the Exercises.
Our main theorem on injective abelian groups (Theorem 3.10) relates them to an apparently very different concept, that of *divisibility*.

DEFINITION 3.8.   An abelian group D is *divisible* if, given any $x \in D$ and any nonzero integer n, there exists $y \in D$ with $ny = x$. In other words, the homomorphism $D \xrightarrow{n} D$, which consists of multiplication by n, is surjective.

Notice that we no not require that division be unique (that is, that $n : D \to D$ be injective). Thus, as examples of divisible abelian groups, we have the rationals **Q**, the reals **R**, the complex numbers **C**, any real or complex vector space; and also **Q/Z**, the rationals mod 1, and **R/Z**, the reals mod 1 or *circle group*. In the case of the last two groups, division is certainly not unique.

We list two important and elementary properties of divisible abelian groups.

PROPOSITION 3.9 D.

(*i*) *A direct sum of divisible abelian groups is divisible.*

(*ii*) *A quotient group of a divisible abelian group (by any subgroup) is divisible.*

PROOF.

(i) If each D_i is divisible and $\{d_i\} \in \oplus D_i$, n a positive integer, we find, for each i, $y_i \in D_i$ with $ny_i = d_i$ and then $n\{y_i\} = \{d_i\}$.

(ii) If $\epsilon : D \twoheadrightarrow E$ with D divisible, $x \in E$, n a positive integer, we find $d \in D$ with $\epsilon d = x$, and then $y \in D$ with $ny = d$. Then $n\epsilon(y) = \epsilon d = x$, so that E is divisible.

We now prove the main theorem.

THEOREM 3.10.   *The abelian group* D *is injective if and only if it is divisible.*

PROOF.   Let D be injective, $x \in D$, n a positive integer. Consider the diagram

$$
\begin{array}{c}
D \\
\uparrow \phi \\
\mathbf{Z} \xleftarrow{\ \ \mu\ \ } \mathbf{Z}
\end{array}
$$

where $\mu(1) = n, \phi(1) = x$. Since D is injective, there exists $\psi : \mathbf{Z} \to D$ with $\psi\mu = \phi$, that is,

$$x = \phi(1) = \psi\mu(1) = \psi(n) = n\psi(1).$$

Thus D is divisible.

Now suppose conversely that D is divisible and we have a diagram

$$D$$
$$\uparrow \phi$$
$$A \overset{\mu}{\longleftarrow\!\!\!<} B$$

We wish to find $\psi : A \to D$ with $\psi\mu = \phi$. For notational convenience we suppose μ to be an embedding, so that B is a subgroup of A, and we wish to extend $\phi : B \to D$ to A. We consider pairs (B_α, ϕ_α), where $B \subseteq B_\alpha \subseteq A$ and $\phi_\alpha : B_\alpha \to D$ is a homomorphism extending ϕ,

$$\phi_\alpha | B = \phi.$$

We order the set W of such pairs by the rule $(B_\alpha, \phi_\alpha) \leqslant (B_\beta, \phi_\beta)$ if

$$B_\alpha \subseteq B_\beta \quad \text{and} \quad \phi_\beta | B_\alpha = \phi_\alpha.$$

Now W is not empty, since $(B, \phi) \in W$, and the ordering on W is *inductive*, that is, given any chain $\{(B_\alpha, \phi_\alpha)\}$ of elements of W, there is an element (B_0, ϕ_0) of W with

$$(B_\alpha, \phi_\alpha) \leqslant (B_0, \phi_0), \tag{3.3}$$

for all (B_α, ϕ_α) in the chain. We merely have to set $B_0 = \cup_\alpha B_\alpha, \phi_0 | B_\alpha = \phi_\alpha$. It is clear that B_0 is a subgroup of A. For if $x \in B_\alpha, y \in B_\beta$ with $(B_\alpha, \phi_\alpha), (B_\beta, \phi_\beta)$ in the chain, then we may suppose $(B_\alpha, \phi_\alpha) \leqslant (B_\beta, \phi_\beta)$, so that $x \in B_\beta, x - y \in B_\beta \subseteq B_0$. Also, by a similar argument, it follows easily that ϕ_0 is well defined and is a homomorphism. Since (3.3) is now obvious, it follows that W is inductively ordered so that, by Zorn's lemma, it has a maximal element $(\bar{B}, \bar{\phi})$. We prove that $\bar{B} = A$, thus completing the proof of the theorem.

We argue by contradiction. If $\bar{B} \neq A$, let $a \in A - \bar{B}$. There are now two possibilities: (a) $na \in \bar{B}$ for no positive integer n; (b) $na \in \bar{B}$ for some positive integer n. In case (a) we set

$$\tilde{\phi}(a) = 0, \quad \tilde{\phi} | \bar{B} = \bar{\phi}, \tag{3.4}$$

and claim we have thus extended $\bar{\phi}$ to the subgroup $\tilde{B}$ of A generated by a and $\bar{B}$. In case (b) let n_0 be the smallest positive integer n such that $na \in \bar{B}$. Since D is divisible we may find $y \in D$ with $n_0 y = \bar{\phi}(n_0 a)$ and we set

$$\tilde{\phi}(a) = y, \quad \tilde{\phi} | \bar{B} = \bar{\phi}. \tag{3.5}$$

Again we claim that (3.5) provides an extension $\tilde{\phi}$ of $\bar{\phi}$ to $\tilde{B}$. We merely have to show that (3.5) is unambiguous. Now if $na \in \bar{B}$, n must be a multiple of n_0; for the set of integers n such that $na \in \bar{B}$ is a subgroup—indeed an ideal—in $\mathbf{Z}$, and it is precisely the subgroup $n_0\mathbf{Z}$. Thus $n = kn_0$, so that $\tilde{\phi}(na) = ny, \bar{\phi}(na) = \bar{\phi}(kn_0 a) = kn_0 y = ny$.

Thus in any case, $(\bar{B}, \bar{\phi}) \leqslant (\tilde{B}, \tilde{\phi})$. But $\tilde{B}$ strictly contains $\bar{B}$, so that we have contradicted the maximality of $(\bar{B}, \bar{\phi})$.

The reader desiring to avoid Zorn's lemma (or transfinite induction) may confine himself to the case when A is countable and use simple induction.

REMARK. The reader may find it a little strange that we listed some elementary properties of injective abelian groups in Proposition 3.2d and some elementary properties of divisible abelian groups in Proposition 3.9—and then proved the two classes of groups to be the same! Our reason is that in Chapter 4 on the theory of modules we exploit both the notions of divisibility and injectivity, but they will no longer coincide in general. Nevertheless, Propositions 3.2(a) and 3.9 will remain true in their more general setting. The situation is similar to that concerning projective abelian groups and free abelian groups.

We are now ready to prove Theorem 3.7.

PROOF OF THEOREM 3.7. Let A be an abelian group and let

$$R \rightarrowtail F \overset{\epsilon}{\twoheadrightarrow} A$$

be a free abelian presentation of A; thus ϵ induces $\bar{\epsilon}: F/R \cong A$. Now $F = \oplus\mathbf{Z}$, the direct sum of copies of $\mathbf{Z}$ indexed by some indexing set S. Let $G = \oplus\mathbf{Q}$, the direct sum of copies of the rationals $\mathbf{Q}$, also indexed by S, and let $F \subseteq G$ be the natural embedding. Then F/R embeds in G/R, so that A embeds in G/R. Since $\mathbf{Q}$ is divisible, so is G (Proposition 3.9(i)): hence so is G/R (Proposition 3.9(ii)). Thus Theorem 3.7 follows from Theorem 3.10.

The reader should now find it easy to establish the equivalence of (i), (iii), and (iv) in Theorem 3.3(a) (using, of course, Theorem 4.1).

4. EXACT SEQUENCES OF ABELIAN GROUPS

Our first theorem in this section serves to simplify the situation described in Section 7 of Chapter 1, relating to direct products and semidirect products (see Exercise 7.3 of Chapter 1 for a more general statement).

Let $0 \rightarrow A' \overset{\mu}{\rightarrow} A \overset{\epsilon}{\rightarrow} A'' \rightarrow 0$ be a s.e.s. of abelian groups.

THEOREM 4.1. *The following statements are equivalent:*

(*i*) *There exists* $\eta : A \to A'$ *with* $\eta\mu = 1$.

(*ii*) *There exists* $\eta : A \to A'$ *so that* $(A; \eta, \epsilon)$ *is the direct product of* A′ *and* A″.

(*iii*) *There exists* $\nu : A'' \to A$ *with* $\epsilon\nu = 1$.

(*iv*) *There exists* $\nu : A'' \to A$ *so that* $(A; \mu, \nu)$ *is the direct sum of* A′ *and* A″.

(*v*) *There exist* $\eta : A \to A', \nu : A'' \to A$ *with* $\mu\eta + \nu\epsilon = 1$.

PROOF. (i)⟺(ii) by Theorem 7.4 of Chapter 1. Plainly (ii)⟹(iii), (iv), (v).

Now, given (iii), consider $1 - \nu\epsilon : A \to A$. Then $\epsilon(1 - \nu\epsilon) = \epsilon - \epsilon\nu\epsilon = \epsilon - \epsilon$ $= 0$, so that $1 - \nu\epsilon = \mu\eta$ for some $\eta : A \to A'$. Thus $1 = \mu\eta + \nu\epsilon$, so that (iii)⟹(v). But then

$$\mu = \mu\eta\mu + \nu\epsilon\mu = \mu\eta\mu,$$

hence, μ being injective, $1 = \eta\mu$. Thus (iii)⟹(v)⟹(i). Since, obviously, (iv)⟹(i) (say), the proof is complete. The reader should note that we use commutativity here precisely in discussing the *homomorphism* $1 - \nu\epsilon$. He should also observe that the implication (iii)⟹(ii) is the ingredient that completes the proofs of Theorem 3.3 and 3.3d (omitting (ii) in the latter).

Our next two results are of great significance in homological algebra, and will be considerably referred to in Chapter 4. They partake of the "duality" to which we have already made reference and which will be clarified in Chapter 3.

We start, as above, with a s.e.s. of abelian groups

$$0 \to A' \xrightarrow{\mu} A \xrightarrow{\epsilon} A'' \to 0. \tag{4.1}$$

THEOREM 4.2. *If* X *is an abelian group, the sequence*

$$0 \to \mathrm{Hom}\,(X, A') \xrightarrow{\mu_*} \mathrm{Hom}\,(X, A) \xrightarrow{\epsilon_*} \mathrm{Hom}\,(X, A'')$$

is exact. Moreover, ϵ_* *is surjective for all s.e.s.* (4.1) *if and only if* X *is projective.*

THEOREM 4.3. *If* Y *is an abelian group, the sequence*

$$0 \to \mathrm{Hom}\,(A'', Y) \xrightarrow{\epsilon^*} \mathrm{Hom}\,(A, Y) \xrightarrow{\mu^*} \mathrm{Hom}\,(A', Y)$$

is exact. Moreover, μ^* *is surjective for all s.e.s.* (4.1) *if and only if* Y *is injective.*

PROOF OF THEOREM 4.2.   That μ_* is injective is precisely the condition that μ is monic. The exactness at $\mathrm{Hom}(X,A)$ is precisely the universal property (Theorem 4.7 of Chapter 1) characterizing the kernel μ of ϵ. The final assertion is the definition, slightly reformulated, of a projective abelian group.

We leave to the reader the proof of Theorem 4.3.

5. THE TENSOR PRODUCT OF ABELIAN GROUPS

In this section we introduce a fundamental construction. Suppose that we are given two abelian groups A and B. We are going to define a new abelian group $A \otimes B$, the *tensor product* of A and B, by giving a system of generators and relators (see the end of Section 2).

The set of generators is just the Cartesian product $A \times B$. The relators are "words" in the generators, namely,

$$(a_1 + a_2, b) - (a_1, b) - (a_2, b), \quad a_1, a_2 \in A, b \in B,$$

$$(a, b_1 + b_2) - (a, b_1) - (a, b_2), \quad a \in A, b_1, b_2 \in B. \tag{5.1}$$

We write $a \otimes b$ for the element of $A \otimes B$ corresponding to (a, b); then we may express this definition of $A \otimes B$ by saying that it is generated by the elements $a \otimes b$, subject to the relations

$$(a_1 + a_2) \otimes b = a_1 \otimes b + a_2 \otimes b, \quad a_1, a_2 \in A, b \in B,$$

$$a \otimes (b_1 + b_2) = a \otimes b_1 + a \otimes b_2, \quad a \in A, b_1, b_2 \in B. \tag{5.2}$$

We first prove some elementary facts.

PROPOSITION 5.1.   *If* $a \in A, b \in B$ *and* m *is any integer, then*

$$m(a \otimes b) = ma \otimes b = a \otimes mb.$$

In particular, $a \otimes 0 = 0 \otimes b = 0$.

PROOF.   If m is positive, the relation $m(a \otimes b) = ma \otimes b$ follows from the first relation in (5.2) by induction. If $m = 0$, we deduce $0 \otimes b = 0$ by setting $a_2 = 0$ in that relation. We then deduce $-(a \otimes b) = -a \otimes b$ by setting $a_2 = -a_1$ in that relation. Then if m is negative, $m = -n$, we have

$$m(a \otimes b) = n(-(a \otimes b)) = n(-a \otimes b) = n(-a) \otimes b = ma \otimes b.$$

The rest of the proposition follows similarly from the second relation in (5.2).

Now when an abelian group G is given by generators S and relators T a function $f: S \to C$, where C is an abelian group, yields a unique homomorphism $\phi: G \to C$ if and only if the extension of f to the free abelian group $F_S (= \mathbf{Z}^{[S]})$ maps every relator to 0; this is a special case of the universal cokernel property [(4.5) of Chapter 1]. In the case of the tensor product, this means that, given a function $f: A \times B \to C$, then f yields a unique homomorphism $\phi: A \otimes B \to C$ if, and only if,

$$f(a_1 + a_2, b) = f(a_1, b) + f(a_2, b), \quad a_1, a_2 \in A, b \in B,$$

$$f(a, b_1 + b_2) = f(a, b_1) + f(a, b_2), \quad a \in A, b_1, b_2 \in B. \tag{5.3}$$

We have a special case of this if C is itself a tensor product, $C = A_1 \otimes B_1$, and $\alpha: A \to A_1, \beta: B \to B_1$, are homomorphisms. We define $f: A \times B \to A_1 \otimes B_1$ by

$$f(a, b) = \alpha a \otimes \beta b, \quad a \in A, b \in B.$$

Since α, β are homomorphisms, it is obvious from (5.2) that (5.3) are satisfied. We write

$$\alpha \otimes \beta : A \otimes B \to A_1 \otimes B_1$$

for the resulting homomorphism, so that $\alpha \otimes \beta$ is determined by

$$(\alpha \otimes \beta)(a \otimes b) = \alpha a \otimes \beta b, \quad a \in A, b \in B. \tag{5.4}$$

Of particular interest are the cases $\beta = 1: B \to B$, or $\alpha = 1: A \to A$. In the former case we write $\alpha_+ : A \otimes B \to A_1 \otimes B$; in the latter case $\beta_+ : A \otimes B \to A \otimes B_1$. No real ambiguity can be caused thereby, especially because the tensor product is naturally *commutative*:

PROPOSITION 5.2. *There is an isomorphism* $\mathrm{A} \otimes \mathrm{B} \cong \mathrm{B} \otimes \mathrm{A}$, *given by*

$$a \otimes b \mapsto b \otimes a.$$

The tensor product is also naturally *associative*:

PROPOSITION 5.3. *There is an isomorphism* $\mathrm{A} \otimes (\mathrm{B} \otimes \mathrm{C}) \cong (\mathrm{A} \otimes \mathrm{B}) \otimes \mathrm{C}$, *given by*

$$a \otimes (b \otimes c) \mapsto (a \otimes b) \otimes c.$$

We next note the following (functorial) properties of $\alpha \otimes \beta$, which follow immediately from (5.4).

PROPOSITION 5.4.
(i) $(\alpha' \otimes \beta')(\alpha \otimes \beta) = \alpha'\alpha \otimes \beta'\beta$.
(ii) $1 \otimes 1 = 1$.
(iii) $\alpha \otimes 0 = 0 \otimes \beta = 0$.

We leave to the reader not only the proof of this elementary proposition, but also its interpretation. We note the consequences:

PROPOSITION 5.5.
(i) If $\alpha : A \to A_1, \beta : B \to B_1$, then $\alpha \otimes \beta = \beta_\dagger \alpha_\dagger = \alpha_\dagger \beta_\dagger$.
(ii) $(\alpha'\alpha)_\dagger = \alpha'_\dagger \alpha_\dagger, (\beta'\beta)_\dagger = \beta'_\dagger \beta_\dagger$.
(iii) $1_\dagger = 1$.

PROPOSITION 5.6. If $(A; q_i)$ is the direct sum of the abelian groups A_i, then $(A \otimes B; q_{i\dagger})$ is the direct sum of the abelian groups $A_i \otimes B$.

PROOF. Given homomorphisms $\xi_i : A_i \otimes B \to X$, we define $\xi : A \otimes B \to X$ by

$$\xi(\{a_i\} \otimes b) = \sum \xi_i(a_i \otimes b). \tag{5.5}$$

The reader will easily verify that (5.5) does describe a homomorphism, using (5.2). It is moreover clear that $\xi q_{i\dagger} = \xi_i$, and that ξ is the unique homomorphism satisfying those last equations.

We write the conclusion of this proposition as

$$\oplus A_i \otimes B = \oplus (A_i \otimes B); \tag{5.6}$$

the meaning is clear from the statement of the proposition. Of course, there is the analogous assertion

$$A \otimes \oplus B_j = \oplus (A \otimes B_j), \tag{5.7}$$

and (5.6) and (5.7) combine to give

$$\oplus_i A_i \otimes \oplus_j B_j = \oplus_{ij} A_i \otimes B_j. \tag{5.8}$$

EXAMPLE 5.7. We may calculate the tensor product of cyclic groups by the observation that if we are given generators x_i and relators $r_\alpha(x)$ for A, and generators y_j and relators $s_\beta(y)$ for B, then $A \otimes B$ is generated by the elements $x_i \otimes y_j$, with relations

$$r_\alpha(x) \otimes y_j = 0,$$

$$x_i \otimes s_\beta(y) = 0.$$

From this we readily deduce that

$$\mathbf{Z} \otimes \mathbf{Z} \cong \mathbf{Z}, \; \mathbf{Z} \otimes \mathbf{Z}_n \cong \mathbf{Z}_n \otimes \mathbf{Z} \cong \mathbf{Z}_n, \; \mathbf{Z}_m \otimes \mathbf{Z}_n \cong \mathbf{Z}_{gcd(m,n)}. \qquad (5.9)$$

In fact, the first three isomorphisms of (5.9) are merely special cases of the relations

$$A \otimes \mathbf{Z} \cong \mathbf{Z} \otimes A \cong A, \qquad (5.10)$$

so that we will be content to prove the last isomorphism in (5.9). Let $\mathbf{Z}_m$ be generated by a, $\mathbf{Z}_n$ by b. Then $\mathbf{Z}_m \otimes \mathbf{Z}_n$ is generated by $a \otimes b$, subject to the relations

$$m(a \otimes b) = 0,$$

$$n(a \otimes b) = 0.$$

But these two relations are equivalent to the single relation $d(a \otimes b) = 0$, where $d = gcd(m,n)$.

From (5.8) and (5.9) and the structure theorem for finitely generated abelian groups we may calculate the tensor product of any two (indeed, any finite number) of finitely generated abelian groups.

We now turn attention to the behavior of the tensor product with respect to exact sequences. Our principal theorem here is the following.

THEOREM 5.8. *Let* $0 \to A' \overset{\mu}{\to} A \overset{\epsilon}{\to} A'' \to 0$ *be a s.e.s. of abelian groups. Then the sequence*

$$A' \otimes B \overset{\mu_+}{\to} A \otimes B \overset{\epsilon_+}{\to} A'' \otimes B \to 0$$

is exact.

PROOF. Since ϵ is surjective and $\epsilon_+(a \otimes b) = \epsilon a \otimes b$, it is obvious that ϵ_+ is surjective. Likewise it is obvious that $\epsilon_+ \mu_+ = 0$. The subtle point is that $\ker \epsilon_+ \subseteq \operatorname{im} \mu_+$. We will give an *ad hoc* proof of this fact here and a second, much more sophisticated but also more revealing proof, in Chapter 4.

Since $\epsilon_+ \mu_+ = 0$, ϵ_+ induces $\bar{\epsilon} : \operatorname{coker} \mu_+ \to A'' \otimes B$, and we show that $\bar{\epsilon}$ is an isomorphism; this is clearly all we need. If $\xi \in A \otimes B$, we write $\bar{\xi} \in \operatorname{coker} \mu_+$ for the coset of ξ mod $\operatorname{im} \mu_+$. Consider the function $f : A'' \times B \to \operatorname{coker} \mu_+$, given by

$$f(a'', b) = \overline{a \otimes b}, \qquad (5.11)$$

where $\epsilon a = a''$. Then f is well defined, since if $\epsilon a_1 = \epsilon a_2$, then $a_1 - a_2 \in \operatorname{im} \mu$, so that $a_1 \otimes b - a_2 \otimes b \in \operatorname{im} \mu_+$, and

$$\overline{a_1 \otimes b} = \overline{a_2 \otimes b}.$$

We show that f, defined by (5.11), satisfies the bilinearity conditions (5.3). This is obvious so far as linearity in b is concerned; it is hardly less obvious in regard to linearity in a'', for if $\epsilon a_i = a_i'', i = 1, 2$, then

$$f(a_1'' + a_2'', b) = \overline{(a_1 + a_2) \otimes b} = \overline{a_1 \otimes b} + \overline{a_2 \otimes b}.$$

Thus f induces a unique homomorphism $\phi : A'' \otimes B \to \operatorname{coker} \mu_+$, with $\phi(a'' \otimes b) = \overline{a \otimes b}$ where $\epsilon a = a''$. Now $\phi \bar{\epsilon} \, \overline{(a \otimes b)} = \phi(\epsilon a \otimes b) = \overline{a \otimes b}$, so that $\phi \bar{\epsilon} = 1$. Thus $\bar{\epsilon}$ is injective; hence, being surjective, it is an isomorphism.

Notice that we do *not* claim that μ_+ is injective. Indeed, in general, it is not. Consider, for example, the case where the s.e.s. is $0 \to \mathbf{Z} \xrightarrow{2} \mathbf{Z} \to \mathbf{Z}_2 \to 0$ and $B = \mathbf{Z}_2$. Then $A' \otimes B = \mathbf{Z}_2$, $A \otimes B = \mathbf{Z}_2$, and $\mu_+ : \mathbf{Z}_2 \to \mathbf{Z}_2$ is again multiplication by 2. Thus $\mu_+ = 0 : \mathbf{Z}_2 \to \mathbf{Z}_2$. In Chapter 7 we make a systematic study of $\ker \mu_+$.

We close this section by noting certain special cases when μ_+ is injective.

THEOREM 5.9.

  (i) *Given the s.e.s.* $0 \to A' \xrightarrow{\mu} A \xrightarrow{\epsilon} A'' \to 0$, *then* $\mu_+ : A' \otimes B \to A \otimes B$ *is injective for all* B *if* $\mu A'$ *is a direct summand in* A.

  (ii) *Given* B, *then* $\mu_+ : A' \otimes B \to A \otimes B$ *is injective for all s.e.s.* $0 \to A' \xrightarrow{\mu} A \xrightarrow{\epsilon} A'' \to 0$ *if* B *is free abelian.*

PROOF.

  (i) Let $A = \mu A' \oplus \nu A''$. Then [see (5.6)] $A \otimes B = \mu_+(A' \otimes B) \oplus \nu_+(A'' \otimes B)$ and μ_+, ν_+ are injective.

  (ii) Let $B = \mathbf{Z}^{[S]}$. Then, by (5.7), $A' \otimes B = A'^{[S]}$, $A \otimes B = A^{[S]}$ and $\mu_+ : A'^{[S]} \to A^{[S]}$ is just the homomorphism that coincides with μ on each summand A' of $A'^{[S]}$. Obviously, since μ is injective, so is μ_+.

Notice that in neither of the cases above do we claim that the condition given is necessary. We will see, in the exercises, that neither condition is, in fact, necessary; the argument is based on the following important observation.

PROPOSITION 5.10. *Suppose that* $a_i \in A, b_i \in B, i = 1, 2, \ldots, k$, *and* $\sum_i a_i \otimes b_i = 0$ *in* $A \otimes B$. *Then there exist finitely generated subgroups* $A_0 \subseteq A, B_0 \subseteq B$, *such that* $a_i \in A_0, b_i \in B_0, i = 1, 2, \ldots, k$, *and* $\sum_i a_i \otimes b_i = 0$ *in* $A_0 \otimes B_0$.

PROOF. Since $\sum_i a_i \otimes b_i = 0$ in $A \otimes B$, it follows that, in $A \times B$, we may express $\sum_i a_i \times b_i$ as a linear combination of (a finite number of) terms of the form $(c + c', d) - (c, d) - (c', d)$ and (a finite number of) terms of the form $(f, g + g') - (f, g) - (f, g')$. Let A_0 be the subgroup of A generated by the a_i and the c, c', f occurring in these terms; and let B_0 be the subgroup of B generated by the b_i and the d, g, g' occurring in these terms. Then A_0, B_0 are finitely generated and satisfy the conclusion of the proposition.

We content ourselves with examining explicitly the implication of this proposition for the converse of Theorem 5.9(ii). Suppose that B is *locally free*, that is, every finitely generated subgroup of B is free abelian. (By Theorem 2.2 this is equivalent to B being torsion-free.) We want to show that if μ embeds A' in A, then μ_+ embeds $A' \otimes B$ in $A \otimes B$. We regard A' as a subgroup of A and we suppose that $\sum a_i \otimes b_i = 0$ in $A \otimes B, a_i \in A'$. Then, by Proposition 5.10, there is a finitely generated subgroup B_0 of B such that $b_i \in B_0$ and $\sum a_i \otimes b_i = 0$ in $A \otimes B_0$. But B_0 is free abelian by hypothesis so that, by Theorem 5.9(ii), $\sum a_i \otimes b_i = 0$ in $A' \otimes B_0$. But then certainly $\sum a_i \otimes b_i = 0$ in $A' \otimes B$, so that μ_+ is injective as claimed.

The argument continues in Exercise 5.4.

We close this section by introducing a class of subgroups relevant to the proof that the condition given in Theorem 5.9(i) is not necessary. Let A' be a subgroup of A; we say that A' is *pure* in A if, for all positive integers n,

$$A' \cap nA = nA'. \tag{5.12}$$

The reader will readily prove

PROPOSITION 5.11. *A direct summand is pure.*

We also have

PROPOSITION 5.12. *The torsion subgroup of an abelian group is pure.*

PROOF. Let T be the torsion subgroup of A and let $a' \in T, a' = na, a \in A$. Then plainly $a \in T$, so that $a' \in nT$ and $T \cap nA = nT$.

Suppose that we have a s.e.s. of abelian groups

$$0 \to A' \xrightarrow{\mu} A \xrightarrow{\epsilon} A'' \to 0. \tag{5.13}$$

It is convenient to say that μ is pure, or that ϵ is pure, or that the s.e.s. is pure, if $\mu A'$ is pure in A.

THEOREM 5.13. *The following statements about (5.13) are equivalent:*
  (i) *(5.13) is pure.*
  (ii) $\mu_+ : A' \otimes \mathbf{Z}_n \to A \otimes \mathbf{Z}_n$ *is injective for all n.*
  (iii) $\epsilon_* : Hom(\mathbf{Z}_n, A) \to Hom(\mathbf{Z}_n, A'')$ *is surjective for all n.*

PROOF. (i)⟺(ii) Statement (ii) is equivalent to $\mu_+ : A_n' \rightarrow A_n$ injective, where $A_n = A/nA, A_n' = A'/nA'$. It is then obvious that the injectivity of μ_+ is equivalent to (5.12).

(i)⟺(iii) Statement (iii) is equivalent to $\epsilon_* : {}_nA \rightarrow {}_nA''$ surjective, where ${}_nA = \{a \in A | na = 0\}, {}_nA'' = \{a'' \in A'' | na'' = 0\}$. Now suppose (5.13) to be pure and let $a'' \in {}_nA''$. Then $a'' = \epsilon(a), a \in A$, and $na = \mu a', a' \in A'$. Since μ is pure, $a' = na_1', a_1' \in A'$, so that $a'' = \epsilon(a - \mu a_1')$ with $n(a - \mu a_1') = 0$. This shows that ϵ_* is surjective. Conversely, suppose that ϵ_* is surjective, and let $na = \mu a', a \in A, a' \in A'$. Then $n\epsilon(a) = 0, \epsilon(a) \in {}_nA''$, so that $\epsilon(a) = \epsilon(a_1), a_1 \in {}_nA$. It follows that $a - a_1 = \mu a_1', a_1' \in A'$, and $\mu a' = na = n(a - a_1) = \mu na_1'$, so that $a' = na_1'$, and μ is pure.

We immediately infer the following "generalization."

THEOREM 5.14. *The following statements about* (5.13) *are equivalent*:

(i) (5.13) *is pure.*

(ii) $\mu_+ : A' \otimes B \rightarrow A \otimes B$ *is injective for all* B.

(iii) $\epsilon_* : Hom(X, A) \rightarrow Hom(X, A'')$ *is surjective for all* X *which are direct sums of cyclic groups.*

PROOF. Of course, in the light of Theorem 5.13, we have only to prove that (i)⟹(ii), (i)⟹(iii).

(i)⟹(ii) From the structure theorem for finitely generated abelian groups and (5.7) we readily infer that if (5.13) is pure, then (ii) holds if B is finitely generated. The general case now follows from Proposition 5.10.

(i)⟹(iii) This follows immediately from Theorem 5.13 and the result of Exercise 5.2.

We thus see that for the the conclusion of Theorem 5.9(i) to hold it is necessary and sufficient that μ be pure. Hence to show that the condition that $\mu A'$ be a direct summand in A is not necessary, we must demonstrate that the converse of Proposition 5.11 is false. We will lead the reader through such a demonstration in the exercises. That it is not trivial to demonstrate this is attested by the following result.

THEOREM 5.15. *Let* (5.13) *be pure and let* A'' *be a direct sum of cyclic groups (for example, let* A'' *be finitely generated). Then* $\mu A'$ *is a direct summand in* A.

PROOF. By Theorem 5.14, we infer that $\epsilon_* : Hom(A'', A) \rightarrow Hom(A'', A'')$ is surjective. Thus there exists $\nu : A'' \rightarrow A$ with $\epsilon\nu = 1 : A'' \rightarrow A''$. The theorem follows from Theorem 4.1.

Thus to find a pure sequence that does not split we must ensure that A'' is not finitely generated. In fact, if A'' is a torsion group, we must ensure that the orders of its elements are unbounded, since one may show that if there exists a positive integer k such that $kA'' = 0$, then A'' is the direct sum of cyclic groups.

EXERCISES

1.1. Compute the group $\text{Hom}(\mathbf{Z}_m, \mathbf{Z}_n)$. Show that $\text{Hom}(\mathbf{Z}, K) \cong K$.

1.2. Show that, if A, B, C are commutative groups, then

$$\text{Hom}(A \times B, C) \cong \text{Hom}(A, C) \times \text{Hom}(B, C),$$

$$\text{Hom}(A, B \times C) \cong \text{Hom}(A, B) \times \text{Hom}(A, C).$$

1.3. Interpreting $\text{Hom}(A, B)$ merely as a *set*, are the isomorphisms of Exercise 1.2 valid for arbitrary groups?

1.4. Show that the rule (1.1) induces a commutative group structure into the set $\text{Func}(G, K)$ of functions from G to K, under which $\text{Hom}(G, K)$ becomes a subgroup of $\text{Func}(G, K)$.

1.5. Referring to definitions (1.4) and (1.5), interpret and prove the relations

$$(\alpha_1 \alpha_2)^* = \alpha_2^* \alpha_1^*,$$

$$(\beta_1 \beta_2)_* = \beta_{1*} \beta_{2*},$$

$$\alpha^* \beta_* = \beta_* \alpha^*$$

2.1. Prove that (a) $\text{Hom}(A, \Pi B_i) \cong \Pi \text{Hom}(A, B_i)$, (b) $\text{Hom}(\oplus A_i, B) \cong \Pi \text{Hom}(A_i, B)$.

2.2. Show that if $\{A_i\}$ is an infinite family of torsion groups, then $\oplus A_i$ is a torsion group, but ΠA_i need not be a torsion group.

2.3. Show that if each A_i is free abelian, then $\oplus A_i$ is free abelian.

2.4. Consider the short exact sequence $A' \rightarrowtail A \twoheadrightarrow A''$ of abelian groups. Let C be a class of abelian groups. Show that

$$A', A'' \in C \Leftrightarrow A \in C$$

if C is (a) the class of finitely generated abelian groups, (b) the class of finite abelian groups, (c) the class of torsion abelian groups, and (d) the class of p-groups.

2.5. Show that the group of rationals $\mathbf{Q}$ is not free abelian.

2.6. Find the rank and invariants $n_i(p)$ for the abelian groups given below in terms of generators and relations.

(a) Generator a, b, c; relations $2a + 2b + 3c = 0, 5a + 2b - 3c = 0$.

(b) Generators a, b, c; relations $2a = 5b, 2b = 5c, 2c = 5a$.

(c) Generators a, b; relations $12b = 0, 6a = 15b$.

2.7. Use the Structure Theorem for Finitely Generated Abelian Groups to prove the following theorem.

THEOREM. *Let* T *be a finite abelian group. Then* $T \cong \bigoplus_{i=1}^{m} \mathbf{Z}_{n_i}$, *where* $n_1 | n_2 | n_3 \ldots | n_m$. *Moreover, the integers* $n_1, n_2, \ldots, n_m$ *are uniquely determined by* T.

3.1. Prove Theorem 3.1.

3.2. Show that (by contrast with Theorem 3.6) a subgroup of a free group F of finite rank may have infinite rank. [*Hint*: take F free of rank $2, R = [F, F]$. This shows that the subgroup may even be normal.]

3.3. Show that if F is free abelian of finite rank, then $F \cong \text{Hom}(F, \mathbf{Z})$. Show that if F is free abelian of infinite rank, then $F \not\cong \text{Hom}(F, \mathbf{Z})$. [*Hint*: In the infinite case, compute cardinalities. In fact, $\text{Hom}(F, \mathbf{Z})$ is not free if F has infinite rank [8]].

3.4. Prove Proposition 3.2d.

3.5. Show that if F is free and D injective, then $\text{Hom}(F, D)$ is injective. [*Hint*: Use Exercise 2.1(b)].

3.6. Write $\mathbf{Q}_1$ for $\mathbf{Q}/\mathbf{Z}$ and call $\text{Hom}(F, \mathbf{Q}_1)$ *cofree* if F is free. Show that a cofree abelian group is injective. Given any abelian group A, let $A^* = \text{Hom}(A, \mathbf{Q}_1)$. Let $\iota : A \to A^{**}$ be the function given by $\iota(a) = \alpha$, where $\alpha(\phi) = \phi(a), \phi \in A^*$. Show that ι is a monomorphism. [*Hint*: Show that if $a \neq 0$, there exists ϕ with $\phi(a) \neq 0$.]

3.7. Show that $\alpha^* : A_2^* \to A_1^*$ is monic (epic) if and only if $\alpha : A_1 \to A_2$ is epic (monic).

3.8. Let $\epsilon : F \twoheadrightarrow A^*$ be a free abelian presentation of A^*. Show that $\epsilon^* \iota$ embeds A in a cofree abelian group (see Exercises 3.6 and 3.7). Deduce Theorem 3.3d.

3.9. Show that (0) is the only finitely generated divisible abelian group and the only divisible free abelian group.

3.10. Show that D admits *unique* division if and only if D admits the structure of a vector space over $\mathbf{Q}$.

4.1. Call an exact sequence of the form $0 \to A' \xrightarrow{\phi} A \xrightarrow{\psi} A''$ *left exact*. Show that

$$0 \to \text{Hom}(X, A') \xrightarrow{\phi_*} \text{Hom}(X, A) \xrightarrow{\psi_*} \text{Hom}(X, A'')$$

is then left exact.

4.2. Dualize Exercise 4.1 by defining a right exact sequence and showing that the induced sequence

$$0 \to \text{Hom}(A'', Y) \xrightarrow{\psi^*} \text{Hom}(A, Y) \xrightarrow{\phi^*} \text{Hom}(A', Y)$$

is then left exact.

4.3. Give examples to show that ϵ_* in Theorem 4.2 and μ^* in Theorem 4.3 may fail to be epic.

4.4. Show that ϵ_* in Theorem 4.2 is epic and μ^* in Theorem 4.3 is epic if (4.1) splits. Show further that then the sequences of these theorems also split.

5.1. Prove Proposition 5.4.

5.2. Give a careful proof of the principle enunciated in Example 5.7 with a view to the determinations (5.9).

5.3. Show that the abelian group B is locally free if and only if B is torsion-free.

5.4. Show that $\mathbf{Q}$ is locally free but not free. Show that, in the notation of Theorem 5.9, μ_+ is injective for all s.e.s. $0 \to A' \to A \to A'' \to 0$ if and only if B is locally free.

5.5. Prove Proposition 5.11.

5.6. Let $A = \prod_n \mathbf{Z}_{p^{2n}}, p$ prime, and let T be the torsion subgroup of A. Show that, if η_n generates $\mathbf{Z}_{p^{2n}}$, then $\{p^n \eta_n\} \notin T$. Let $b \in A/T$ be the coset of $\{p^n \eta_n\}$. Show that b is divisible in A/T by every nonzero integer, and infer that A/T is not isomorphic to a subgroup of A. Deduce that T is not a direct summand in A; hence that pure subgroups may fail to be direct summands.

5.7. (Harder) Show that if p is prime and $pA = 0$, then $A \cong \oplus \mathbf{Z}_p$, a direct sum of copies of $\mathbf{Z}_p$. Hence show by induction on n that if $p^n A = 0$, then A is a direct sum of cyclic groups. [*Hint*: Express pA as a direct sum of cyclic groups $\mathbf{Z}_{p^{m_\alpha}}$, generated by pa_α. Extend a basis of $_pA \cap pA$ by elements a_β to a basis of $_pA$. Show that A is the direct sum of cyclic groups of order $p^{m_\alpha + 1}$, generated by a_α, and cyclic groups of order p, generated by a_β.] Here
$$mA = \{a \in A \mid a = ma', a' \in A\},$$
$$_mA = \{a \in A \mid ma = 0\}.$$

3

CATEGORIES AND FUNCTORS

In Chapters 1 and 2 we were implicitly considering two important algebraic *categories*—the category of groups and homomorphisms and the category of abelian groups and homomorphisms. We considered certain structural features of these categories (such as the existence of kernels and cokernels, exact sequences, free groups, and free abelian groups) and certain universal constructions (such as free products, direct products, and direct sums). Moreover, we made a fairly detailed comparison of the two categories.

In this chapter we make the language of categories and functors quite explicit in order to discuss the two categories already referred to, together with others, with proper mathematical precision. We will then also have a language for comparing the features of different categories, examining their similarities and differences. The language will be applied systematically in subsequent chapters.

We should explain that we have delayed establishing this language till now because the effective use of language must follow, not precede, the acquisition of familiarity with concepts that the language is designed to formalize. The reader is encouraged to review Chapters 1 and 2 and to recast the various notions there, not explicitly discussed in this chapter, in the new language.

Since our emphasis is very much on language rather than on the development of category *theory*, we include in this chapter only those parts of the theory that find application in this book. The reader is referred to [6] for a comprehensive introduction to category theory.

1. CATEGORIES

We first give the formal definition of a category and then provide some examples. To define a *category* $\mathfrak{C}$ we must give three pieces of data:

(i) A class of *objects* $A, B, C, \ldots$,

(ii) To each pair of objects A, B of $\mathfrak{C}$, a set $\mathfrak{C}(A, B)$ of *morphisms from A to B*,

(iii) To each triple of objects A, B, C of $\mathfrak{C}$, a *law of composition*

$$\mathfrak{C}(A, B) \times \mathfrak{C}(B, C) \to \mathfrak{C}(A, C).$$

Before giving the axioms that a category must satisfy we introduce some auxiliary notation and terminology. If $f \in \mathfrak{C}(A, B)$ we may think of the morphism f as a generalized "function" from the *domain* A to the *range* (or *codomain*) B and write

$$f : A \to B \quad \text{or} \quad A \xrightarrow{f} B.$$

The set $\mathfrak{C}(A, B) \times \mathfrak{C}(B, C)$ consists, of course, of pairs (f, g) where $f : A \to B, g : B \to C$, and we write the composition of f and g as $g \circ f$, or simply gf. The rationale for this notation lies in the fact that if A, B, C are sets and f, g are functions, the composite function from A to C is the function h given by

$$h(a) = g(f(a)), \quad a \in A.$$

Thus if the function symbol is written, as usual, to the *left* of the argument symbol, one is naturally led to write $h = gf$. Of course, it will turn out that sets, functions, and function composition do, in fact, constitute a category, so that it is sensible, if not necessarily optimal, to adopt notations suggested by set theory.

We are now ready to state the axioms for a category. The first one is really more of a convention, the two others being much more substantial.

A_1: The sets $\mathfrak{C}(A_1, B_1), \mathfrak{C}(A_2, B_2)$ are disjoint unless $A_1 = A_2, B_1 = B_2$.

A_2: Given $f : A \to B, g : B \to C, h : C \to D$, then $h(gf) = (hg)f$ (*Associative law of composition*).

A_3: To each object A there is a morphism $1_A : A \to A$ such that, for any $f : A \to B, g : C \to A, f 1_A = f, 1_A g = g$ (*Existence of identities*).

It is easy to see that the morphism 1_A is uniquely determined by axiom $A3$. We call 1_A the *identity morphism* of A, and we often suppress the suffix A, writing simply

$$f1 = f, \quad 1g = g.$$

As remarked, and readily verified, the collection $\mathfrak{S}$ of sets, functions, and function composition satisfies the axioms. We often refer to the *category of sets* $\mathfrak{S}$; indeed, more generally, in describing a category we habitually omit reference to the law of composition when the morphisms are functions and composition is ordinary function composition (or when, for some other reason, the law of composition is evident), and we even omit reference to the nature of the morphisms if the context, or custom, also makes their nature obvious.

Notice that axiom A1 insists that we distinguish two morphisms unless their domains *and ranges* coincide. The reader should also note that the composition gf is only defined if the range f *coincides* with the domain of g. Both of these conventions are frequently violated in classical mathematical usage.

We say that a morphism $f : A \to B$ in $\mathfrak{C}$ is an *isomorphism* (or *invertible* or a *unit*) if there exists a morphism $g : B \to A$ in $\mathfrak{C}$ such that

$$gf = 1_A, \quad fg = 1_B.$$

It is plain that g is then itself invertible and is uniquely determined by f; we write $g = f^{-1}$, so that

$$(f^{-1})^{-1} = f.$$

It is also plain that the composite of two invertible morphisms is again invertible and thus the relation that declares that $A \sim B$ if there exists an invertible morphism $f : A \to B$ (we say "A is *isomorphic* to B" or "A is *equivalent* to B") is an equivalence relation on the objects of the category $\mathfrak{C}$. This relation has special names in different categories (one-one correspondence of sets, isomorphism of groups, homeomorphism of topological spaces), but it is important to observe that it is a *categorical* concept.

We now list several examples of categories.

(a) The category $\mathfrak{S}$ of sets and functions.

(b) The category $\mathfrak{X}$ of topological spaces and continuous functions.

(c) The category $\mathfrak{G}$ of groups and homomorphisms.

(d) The category $\mathfrak{Ab}$ of abelian groups and homomorphisms.

(e) The category $\mathfrak{D}_K$ of vector spaces over the field K and linear transformations.

(f) The category $\mathfrak{S}_0$ of pointed sets (that is, sets with distinguished element) and pointed functions (that is, functions sending distinguished element to distinguished element).

Plainly the list could be continued indefinitely. Plainly also each category carries its appropriate notion of invertible morphisms and isomorphic

CATEGORIES 67

objects. In all the examples given the morphisms are structure-preserving *functions*; however, it is important to emphasize that the morphisms of a category need not be functions, even when the objects of the category are sets perhaps with additional structure. Topology provides us with many such examples, but we will here be content to show how two very familiar examples of structured sets may themselves be regarded as categories.

Suppose G is a group; we form a category $\mathfrak{C}_G$. Then $\mathfrak{C}_G$ has just one object, which we will call 0. We define $\mathfrak{C}_G(0,0)=G$, that is, the morphism set is just the underlying set of G. Composition of morphisms is the group composition. Axiom A_1 is satisfied trivially and axioms A_2 and A_3 reflect the associativity of the group operation and the existence of a unity element in a group. Since every element of a group is invertible, we see that every morphism of $\mathfrak{C}_G$ is invertible. A group is thus a category with one object, every morphism of which is a unit.

Suppose that S is a preordered set. That is, S is a set with a relation $\leqslant$ that is reflexive: $x \leqslant x$ for every x; and transitive: if $x \leqslant y$ and $y \leqslant z$, then $x \leqslant z$. We do not insist on the antisymmetry condition. We form a category $\mathfrak{C}_S$. First, the objects of $\mathfrak{C}_S$ are the elements of S. Second, $\mathfrak{C}_S(x,y)=$ the singleton (x,y) if $x \leqslant y$; it is $\varnothing$ otherwise.

Since $\mathfrak{C}_S(x,y)$ is empty or has only one element, there is only one possible way to compose morphisms. Axiom A_1 holds by virtue of the nature of the singleton $\mathfrak{C}_S(x,y)$. Axiom A_2 reflects the transitive property. Axiom A_3 reflects the reflexive property. A preordered set is thus a category whose morphism sets are singletons or the empty set. Notice that two objects x,y are equivalent if and only if $x \leqslant y$ and $y \leqslant x$. Thus if S is an ordered set (that is, if we require the antisymmetry condition), isomorphic objects of $\mathfrak{C}_S$ are identical.

DEFINITION 1.1. An object I of $\mathfrak{C}$ is *initial* if $\mathfrak{C}(I,X)$ is a singleton for all $X \in \mathfrak{C}$. An object T is *terminal* (or *coinitial*) if $\mathfrak{C}(X,T)$ is a singleton for all $X \in \mathfrak{C}$. An object Z is a *zero object* if it is both initial and terminal.

Suppose now that I and J are both initial objects of $\mathfrak{C}$ and that $\mathfrak{C}(I,J)=\{f\}$, $\mathfrak{C}(J,I)=\{g\}$. We also know that $\mathfrak{C}(I,I)=\{1\}$ and $\mathfrak{C}(J,J)=\{1\}$. We thus have the following diagrams.

$$I \xrightarrow{f} J \xrightarrow{g} I \qquad J \xrightarrow{g} I \xrightarrow{f} J \qquad gf=1,\ fg=1$$

They show that I and J are equivalent. In fact, they show that they are equivalent in only one way. That is, there is one and only one isomorphism from I to J. We thus say that any two initial objects in a category are *canonically* equivalent. The same reasoning shows that terminal objects are canonically equivalent. If $\mathfrak{C}$ has a zero object, then every initial object and every terminal object is a zero object.

EXAMPLES.
1. In $\mathfrak{S}$ an initial object is $\varnothing$ and any singleton is a terminal object. $\mathfrak{S}$ has no zero objects, since $\varnothing$ is not equivalent to a singleton in $\mathfrak{S}$.
2. In $\mathfrak{G}$ a group with one element is a zero object.

Suppose that $A, B \in \mathfrak{C}$ and Z, Z' are zero objects. We have the following diagram, which commutes since Z and Z' are zero objects.

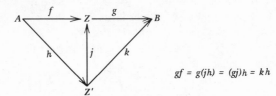

$$gf = g(jh) = (gj)h = kh$$

We have thus picked out a distinguished morphism from A to B by going through a zero object, and it does not matter through which zero object we go. This morphism is called the *zero morphism* from A to B and is denoted 0_{AB}, or simply 0.

PROPOSITION 1.1.
$$0f = 0, \quad g0 = 0.$$

2. FUNCTORS

We begin by introducing the most fundamental concept of category theory.

DEFINITION 2.1. If $\mathfrak{C}$ and $\mathfrak{D}$ are categories, a *functor* $F: \mathfrak{C} \to \mathfrak{D}$ is a rule associating with each object X in $\mathfrak{C}$ an object FX in $\mathfrak{D}$, and with each morphism $f: X \to Y$ in $\mathfrak{C}$ a morphism $Ff: FX \to FY$ in $\mathfrak{D}$ such that

$$F(fg) = (Ff)(Fg)$$

$$F1_X = 1_{FX}.$$

Functors are essentially morphisms of categories.

EXAMPLES.
1. If we consider groups as (one-object) categories, functors are precisely homomorphisms.
2. If we consider preordered sets as categories, functors are precisely order-preserving functions.

3. We next describe a functor from $\mathfrak{G}$ to $\mathfrak{G}$. Suppose G is a group. Let G' be the subgroup generated by $\{xyx^{-1}y^{-1} : x,y \in G\}$. Then G' is a normal subgroup called the *commutator subgroup* of G. The quotient group G/G' is an abelian group, called the *abelianization* of G. We now define a functor Abel: $\mathfrak{G} \to \mathfrak{G}$. If $G \in \mathfrak{G}$, Abel $(G) = G/G'$, where G' is the commutator subgroup. If $f: G \to H$ is a homomorphism, then $g = $ Abel (f): $G/G' \to H/H'$ is the induced homorphism:

$$g(xG') = (fx)H', \quad x \in G.$$

This definition makes good sense, since $fG' \subseteq H'$, and it is plain that Abel is a functor.

Let us consider a specific instance of this abelianizing functor. P_3 is the symmetric group on three elements (see Example 3.5.3 of Chapter 1). Now $(12) = (12)^{-1}$ and $(13) = (13)^{-1}$. Also

$$(12)(13)(12)(13) = (132).$$

Thus the commutator subgroup of P_3 is $\mathbf{Z}_3$, the cyclic group of three elements generated by (123), and

$$\text{Abel}(P_3) = P_3/\mathbf{Z}_3 = \mathbf{Z}_2.$$

Consider the two morphisms in $\mathfrak{G}$:

$$\mathbf{Z}_3 \xrightarrow[i]{} P_3 \qquad \mathbf{Z}_3 \xrightarrow[1]{} \mathbf{Z}_3;$$
$$\text{(embedding)} \qquad \text{(identity)}$$

Abelianizing the morphisms above, we get:

$$\mathbf{Z}_3 \xrightarrow{0} \mathbf{Z}_2 \qquad \mathbf{Z}_3 \xrightarrow{1} \mathbf{Z}_3$$

This shows the importance of taking the view (see Axiom $A1$ for a category) that the homomorphisms i, 1 above, though they are defined on the same domain $\mathbf{Z}_3$ and take the same values at each element of the domain, are different. They are distinguished by having different codomains.

In the same spirit, it is possible to define a functor $\overline{\text{Abel}}: \mathfrak{G} \to \mathfrak{A}\mathfrak{b}$, different from Abel, because the codomain category is different. We will see later (see Exercise 5.7) the importance of this distinction.

4. Suppose that $A \in \mathfrak{S}, A \neq \varnothing$. We define a functor $-^A : \mathfrak{G} \to \mathfrak{G}$. If $G \in \mathfrak{G}$, $-^A(G) = G^A$, the set of functions from A to G, regarded as a group via the group operation in G.

If $f : G \to H$ is a homomorphism, $f^A : G^A \to H^A$ is the function $f^A(g) = fg$, for $g \in G^A$. We have the following diagram to describe the functor.

$$
\begin{array}{ccc}
G & \longmapsto & G^A \\
f\downarrow & & \downarrow f^A \\
H & \longmapsto & H^A
\end{array}
$$

Here the arrow $\longmapsto$ indicates the value of the functor on the objects G, H. This notation will be used consistently, for functions as well as for functors. It is obvious that $-^A$ is a functor.

5. *The covariant power set functor.* P. We define a functor $P : \mathfrak{S} - \mathfrak{S}$ as follows. If $S \in \mathfrak{S}$, $PS = 2^S$, the collection of subsets of S. If $f : X \to Y, Pf = 2^f$ where $2^f(A) = fA, A \subseteq X$. Again we have

$$
\begin{array}{ccc}
X & \longmapsto & 2^X \\
f\downarrow & & \downarrow 2^f \\
Y & \longmapsto & 2^Y
\end{array}
$$

It is obvious that P is a functor.

6. *The differentiation functor.* Let $\mathfrak{D}$ be the category whose objects are pairs (I, x_0), where I is an open interval of the real line containing x_0. A morphism $f : (I, x_0) \to (J, y_0)$ is then a differentiable function $f : I \to J$ such that $f(x_0) = y_0$. Let $\mathrm{Hom}(\mathbf{R}, \mathbf{R})$ be the category with one object, whose morphisms are linear maps $\mathbf{R} \to \mathbf{R}$. We define a functor $D : \mathfrak{D} \to \mathrm{Hom}(\mathbf{R}, \mathbf{R})$ by the rule (in traditional notation)

$$(Df)(x) = y_0 + f'(x_0)(x - x_0), \quad x \in \mathbf{R},$$

The reader will check that the functorial property $D(f \circ g) = Df \circ Dg$ is just the familiar "chain rule"

$$\frac{dz}{dx} = \frac{dz}{dy}\frac{dy}{dx}$$

We may generalize this example to functions of several variables.

We now proceed to define two very important classes of functors.

7. *Underlying (or forgetful) functors.* An example of an *underlying* functor

is $U: \mathfrak{G} \to \mathfrak{S}$. Here $U(G) = $ the set underlying the group G (without the binary operation); and $U(f)$ is the function f (not considered as a homomorphism.) We will later explain why underlying functors are important in category theory. Here we merely record one notational convenience to which they give rise. Suppose that N is a normal subgroup of G and $f: G \to H$ is a homomorphism such that $fN = 1$ (that is, $N \subseteq$ kernel f). We have the following diagram.

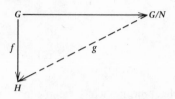

We can prove the existence of a homomorphism g that causes the diagram to commute. But what is g before we prove it is a homomorphism? It is simply a morphism of $\mathfrak{S}$, that is, a function. The diagram above is transferred to $\mathfrak{S}$ by the underlying functor and then brought back to $\mathfrak{G}$ by a proof that g is in the image of U. Naturally we do not insist, in ordinary mathematical discourse, on mentioning underlying functors. Other examples of underlying functors are:

$$U: \mathfrak{D}_K \to \mathfrak{A}\mathfrak{b}$$

$$U: \mathfrak{D}_K \to \mathfrak{S}_K, \quad \text{sets with scalar multiplication,}$$

$$U: \mathfrak{T} \to \mathfrak{S}.$$

DEFINITION 2.2. Let $F: \mathfrak{C} \to \mathfrak{D}$ be a functor and $X, Y \in \mathfrak{C}$. Then F induces a function $F_*: \mathfrak{C}(X, Y) \to \mathfrak{D}(FX, FY)$. F is *faithful* if every such induced function is injective. F is *full* if every such induced function is surjective.

All of the underlying functors mentioned above are faithful.

8. *Free functors.* Suppose that $S \in \mathfrak{S}$. Denote by $\mathrm{Fr}(S)$ the free abelian group on S as basis. (If $S = \varnothing, \mathrm{Fr}(S) = \{0\}$.) If $f: S \to T$, $\mathrm{Fr}(f): \mathrm{Fr}(S) \to \mathrm{Fr}(T)$ is the homomorphism determined on the basis S by $\mathrm{Fr}(f)(x) = fx$, for $x \in S$. Then $\mathrm{Fr}: \mathfrak{S} \to \mathfrak{A}\mathfrak{b}$ is a functor, called the *free functor from $\mathfrak{S}$ to $\mathfrak{A}\mathfrak{b}$*. There are other free functors; for example, from $\mathfrak{S}$ to $\mathfrak{G}$, from $\mathfrak{S}$ to $\mathfrak{D}_K$.

DEFINITION 2.3. Let $\mathfrak{C}, \mathfrak{D}$ be categories with zero object. Then a functor $F: \mathfrak{C} \to \mathfrak{D}$ is *zero-preserving* if $F0 = 0$.

3. NATURAL TRANSFORMATIONS

DEFINITION 3.1.   Suppose that $F, G : \mathfrak{C} \to \mathfrak{D}$ are functors. A *natural transformation* $\tau : F \to G$ is a rule assigning to each $X \in \mathfrak{C}$ a morphism $\tau_X : FX \to GX$ in $\mathfrak{D}$ such that for every $f : X \to Y$, the following diagram commutes:

$$
\begin{array}{ccc}
FX & \xrightarrow{\tau_X} & GX \\
{\scriptstyle Ff}\downarrow & & \downarrow{\scriptstyle Gf} \\
FY & \xrightarrow[\tau_Y]{} & GY
\end{array}
$$

The idea of understanding what we mean when we say something is "natural" was the basic idea motivating Eilenberg and MacLane in their development of category theory.

As a fundamental example of a natural transformation we consider the category $\mathfrak{D}_K$. Suppose that $V \in \mathfrak{D}_K$. Then $V^* \in \mathfrak{D}_K$, where V^* is the dual space of all linear transformations from V to K. If V is finite-dimensional, then V is isomorphic to V^*. Similarly we can form V^{**}, and, if V is finite-dimensional, V is isomorphic to V^{**}. Indeed, for any $V \in \mathfrak{D}_K$, let $v \in V$, and define $\tilde{v} \in V^{**}$ as follows:

$$
\tilde{v}(f) = f(v), \quad f \in V^*.
$$

The function $i : v \mapsto \tilde{v}$ is a linear transformation from V to V^{**}, which is an isomorphism if V is finite-dimensional. Furthermore, it is a "natural" isomorphism in that it is constructed without reference to any particular basis, whereas to describe an isomorphism between V and V^* dual bases must be chosen and the isomorphism depends on the choice of bases. The definition above makes precise our intuitive notion that i is natural. Thus let us denote by $\mathrm{Id} : \mathfrak{D}_K \to \mathfrak{D}_K$ the identity functor, and let us denote by $** : \mathfrak{D}_K \to \mathfrak{D}_K$ the functor given by $**(V) = V^{**}$ on objects. If $f : V \to W$, then f^{**} is the linear transformation

$$
f^{**} : V^{**} \to W^{**}
$$

defined by $(f^{**}v^{**})\phi = v^{**}(f^*\phi)$, where $v^{**} : V^* \to K$, $\phi : W \to K$, so that $v^{**} \in V^{**}$, $\phi \in W^*$, and $f^* : W^* \to V^*$ is given by $(f^*\phi)v = \phi(fv)$, $v \in V$. We have a natural transformation

$$
\eta : \mathrm{Id} \to **.
$$

For every $V \in \mathfrak{D}_K, \eta_V : V \to V^{**}$ is the function i described above. The following diagram can be shown to commute:

$$
\begin{array}{ccc}
V & \xrightarrow{\eta_V} & V^{**} \\
{\scriptstyle f}\downarrow & & \downarrow{\scriptstyle f^{**}} \\
W & \xrightarrow[\eta_W]{} & W^{**}
\end{array}
$$

In the case where we consider Id and ** as functors on the category of finite-dimensional vector spaces, $\eta : \text{Id} \to ^{**}$ is a *natural equivalence*, according to the following definition.

DEFINITION 3.2. A natural transformation $\tau : \mathfrak{C} \to \mathfrak{D}$ is a *natural equivalence* if, for every object X in $\mathfrak{C}$, τ_X is a unit.

Another way of putting this is that τ has an inverse. This makes sense as we can compose natural transformations. Thus suppose that $\tau : F \to G$ and $\sigma : G \to H$. Then $\sigma\tau : F \to H$ is described by the diagram below, which shows that it is again a natural transformation.

$$
\begin{array}{ccccc}
FX & \xrightarrow{\tau_X} & GX & \xrightarrow{\sigma_X} & HX \\
{\scriptstyle Ff}\downarrow & & {\scriptstyle Gf}\downarrow & & \downarrow{\scriptstyle Hf} \\
FY & \xrightarrow[\tau_Y]{} & GY & \xrightarrow[\sigma_Y]{} & HY
\end{array}
$$

Notice that if each τ_X is a unit, we can define $\sigma : G \to F$ by $\sigma_X = (\tau_X)^{-1}$; σ is then a natural transformation and we may write $\sigma = \tau^{-1}$.

By analogy with function spaces we make the following construction. Suppose $\mathfrak{C}$ is *small*, that is, the collection of objects of $\mathfrak{C}$ forms a set. We can form a category $\mathfrak{D}^{\mathfrak{C}}$ as follows: The objects of $\mathfrak{D}^{\mathfrak{C}}$ are functors $F : \mathfrak{C} \to \mathfrak{D}$, and $\mathfrak{D}^{\mathfrak{C}}(F, G)$ is the set of natural transformations from $\mathfrak{C}$ to $\mathfrak{D}$. Composition in $\mathfrak{D}^{\mathfrak{C}}$ is then the composition of natural transformations described above.

We may also compose functors and natural transformations. Suppose that $\mathfrak{B} \xrightarrow{E} \mathfrak{C} \underset{G}{\overset{F}{\rightrightarrows}} \mathfrak{D} \xrightarrow{H} \mathfrak{E}$ and $\tau : F \to G$ is a natural transformation. We can form new natural transformations

$$
\tau E : FE \to GE
$$

$$
H\tau : HF \to HG.
$$

They are defined as follows:

$$(\tau E)_B = \tau_{EB}$$

$$(H\tau)_X = H(\tau_X)$$

Notice that $(H\tau)E = H(\tau E)$. The reader should prove that this holds and that $\tau E, H\tau$ are natural.

4. DUALITY PRINCIPLE

Given any category $\mathfrak{C}$, we can form a new category $\mathfrak{C}^{op}$, or sometimes $\mathfrak{C}^0$, as follows. The objects of $\mathfrak{C}^{op}$ are the objects of $\mathfrak{C}$, and

$$\mathfrak{C}^{op}(Y,X) = \mathfrak{C}(X,Y).$$

If f and g are morphisms of $\mathfrak{C}^{op}$,

$$fo^{op}g = gof$$

$$Y \xrightarrow{g} X \xrightarrow{f} W \qquad\qquad Y \xleftarrow{g} X \xleftarrow{f} W$$
$$\underset{fo^{op}g}{\longrightarrow} \qquad\qquad\qquad \underset{gof}{\longleftarrow}$$

in $\mathfrak{C}^{op}$ in $\mathfrak{C}$

$\mathfrak{C}^{op}$ is the *opposite category* of $\mathfrak{C}$, and we go from $\mathfrak{C}$ to $\mathfrak{C}^{op}$ by "reversing arrows."

PROPOSITION 4.1.

(i) $\mathfrak{C}^{op}$ *is a category with the same identities as* $\mathfrak{C}$.
(ii) *If* $\mathfrak{C}$ *has zero morphisms, then these are also zero morphisms in* $\mathfrak{C}^{op}$.
(iii) $(\mathfrak{C}^{op})^{op} = \mathfrak{C}$.

To prove (i), note that a left and right identity is $\mathfrak{C}$ is a right and left identity in $\mathfrak{C}^{op}$. To prove (ii), note that $0g = f0 = 0$ in $\mathfrak{C}$ implies $g0 = 0f = 0$ in $\mathfrak{C}^{op}$. (iii) is obvious.

We now proceed to an informal description of the *duality principle*. Suppose that S is a statement meaningful in any category, and let $S(\mathfrak{C})$ be the statement in $\mathfrak{C}$. Then $S(\mathfrak{C}^{op})$ is a meaningful statement in the category $\mathfrak{C}^{op}$. Since the objects and morphisms of $\mathfrak{C}^{op}$ are the same as those of $\mathfrak{C}$, $S(\mathfrak{C}^{op})$ can be interpreted as a statement in $\mathfrak{C}$. Call this statement $S^{op}(\mathfrak{C})$. S^{op} is the *dual* of the statement S.

EXAMPLE.

$S(\mathbb{C})$: I is initial in $\mathbb{C}$ if $\mathbb{C}(I,X)$ is a singleton for every $X \in \mathbb{C}$.

$S(\mathbb{C}^{\mathrm{op}})$: I is initial in $\mathbb{C}^{\mathrm{op}}$ if $\mathbb{C}^{\mathrm{op}}(I,X)$ is a singleton for every $X \in \mathbb{C}^{\mathrm{op}}$.

$S^{\mathrm{op}}(\mathbb{C})$: I is *coinitial* in $\mathbb{C}$ if $\mathbb{C}(X,I)$ is a singleton for every $X \in \mathbb{C}$.

Of course, coinitial objects are terminal objects. Initial and terminal are thus dual concepts. Having already proved that initial objects are essentially unique, we have also proved that terminal objects are essentially unique. This is because the argument for initial objects is true in any category. It is thus true in op-categories. Interpreting the argument in an op-category as a set of statements about the original category, we would have a proof for terminal objects.

We now proceed to a second, more substantial example.

DEFINITION 4.1.   A morphism f is *monic*, or is a *monomorphism*, if $fu = fv$ implies $u = v$ for all morphisms u, v.

$S(\mathbb{C})$: f is monic in $\mathbb{C}$ if $fou = fov$ implies $u = v$.

$S(\mathbb{C}^{\mathrm{op}})$: f is monic in $\mathbb{C}^{\mathrm{op}}$ if $fo^{\mathrm{op}}u = fo^{\mathrm{op}}v$ implies $u = v$.

$S^{\mathrm{op}}(\mathbb{C})$: f is *epic* in $\mathbb{C}$ if $uof = vof$ implies $u = v$.

$S^{\mathrm{op}}(\mathbb{C})$ gives us the definition of *epic*, or *epimorphism.*

Now let us enunciate, to illustrate the principle, a simple proposition.

PROPOSITION 4.2.   *If* gf *is monic so is* f.

PROOF.   If $fu = fv$ then $gfu = gfv$, so that, gf being monic, $u = v$. Thus f is monic.

We also have a dual proposition, which is *already proved* by the argument above. We will henceforth indicate dual propositions, etc., by superscript stars.

PROPOSITION 4.2*.   *If* fg *is epic so is* f.

The general situation here is that every time we formulate a categorical concept we have a dual concept; the same is true of theorems about categories.

We now characterize the monics and epics in $\mathfrak{S}$.

PROPOSITION 4.3.

(*i*) f *is monic in* $\mathfrak{S}$ *if and only if* f *is injective,*

(*ii*) f *is epic in* $\mathfrak{S}$ *if and only if* f *is surjective.*

PROOF.

(i) Suppose that $f : X \to Y$ is injective and $fu = fv$, with $u, v : W \to X$. Then, for every $w \in W$, $fu(w) = fv(w)$, so that, f being injective, $u(w) = v(w)$. Thus $u = v$ and f is monic.

Conversely, suppose that f is monic and $f(x_1) = f(x_2)$. Let (0) be the singleton set and define $u, v: (0) \to X$ by $u(0) = x_1, v(0) = x_2$. Then $fu = fv$, so that, f being monic, $u = v$. But this means that $x_1 = x_2$, so that f is injective.

(ii) Suppose that $f: X \to Y$ is surjective and $uf = vf$. Then $uf(x) = vf(x)$ for every x in X. Since f is surjective, for every $y \in Y$, there exists $x \in X$ such that $f(x) = y$. Thus $u(y) = v(y)$ for every $y \in Y$ and $u = v$. Therefore f is epic.

Next suppose that f is nonsurjective. then there exists $y_0 \in Y - fX$. Let $Z = \{0, 1\}$. Define

$$u: Y \to Z \text{ by } u(y) = 0, \text{ all } y$$

$$v: Y \to Z \text{ by } v(y) = 0, \text{ if } y \neq y_0 \text{ and } v(y_0) = 1.$$

Then $u \neq v$, but $uf = vf$ since y_0 is not in the image of f. Thus f is nonepic.

Specializing Propositions 4.2, 4.2* to the category of sets we have the following results. These results are quite trivial, of course, and very easily proved without reference to categorical concepts. Here we are concerned with them in relation to categorical duality.

PROPOSITION 4.4.

(i) If gf *is injective then so is* f.

(ii) If fg *is surjective then so is* f.

These two propositions are not strictly dual, although we often loosely describe them so. We can form duals at the categorical level, but not in a specific category.

Notation. We will henceforth denote a monomorphism by $A \overset{f}{\rightarrowtail} B$ and an epimorphism by $A \overset{f}{\twoheadrightarrow} B$.

We now give another example of the duality principle.

Recall that, if $f \in \mathscr{G}(G, H)$, the kernel of f is $K = \{x \in G : fx = e\}$, e being the identity of H. We write $K = f^{-1}(e)$. In categories we do not have elements of sets to work with and therefore need a characterization in terms of morphisms. Thus we must first replace the statement that K is a subgroup of G by the explicit embedding $K \overset{\kappa}{\rightarrowtail} G \overset{f}{\to} H$. Note:

(i) $f\kappa = 0$ (0 is the constant homomorphism). Also κ is somehow "maximal" for this property of being annihilated by composing on the left with f. We make this precise below.

(ii) Suppose that $fg = 0$. Then $g = \kappa h$ for some h.

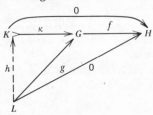

It can be proved that properties (i) and (ii) characterize the kernel (up to canonical isomorphism). However, it is more efficient to postpone this proof until we have given the definition of a kernel in *any* category; this definition should by now be obvious.

[From here on, diagrams (like the one above) are always commutative. Solid arrows are part of the data and dotted arrows are morphisms whose existence is given by a definition or will be proved.]

DEFINITION 4.2. Let $\mathfrak{C}$ be a category with zero, and suppose that $f \in \mathfrak{C}(G, H)$. Then $\kappa : K \to G$ is a *kernel* of f if κ is monic and
(i) $f\kappa = 0$,
(ii) whenever $fg = 0$, we have $g = \kappa h$ for some h.
The diagram is the one above. We leave the proof of the uniqueness of the kernel to the reader.

In Definition 4.2 we could equally well leave out the requirement that κ be monic, and instead assert that h is unique.

Now the dual of Definition 4.2 gives us the *cokernel*.

DEFINITION 4.2*. Let $\mathfrak{C}$ be a category with zero, and suppose that $f \in \mathfrak{C}(H, G)$. Then $\kappa : G \to K$ is a *cokernel* of f if κ is epic and
(i) $\kappa f = 0$,
(ii) whenever $gf = 0$, we have $g = h\kappa$ for some h.

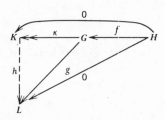

Thus κ is the cokernel of f in $\mathfrak{C}$ if and only if κ is the kernel of f in $\mathfrak{C}^{\mathrm{op}}$.

We now prove an elementary result about kernels in order to exemplify the way we reason at the categorical level.

PROPOSITION 4.6. *Suppose that $\mathfrak{C}$ is a category with zero. Then every monic has a kernel; namely the kernel of* $f : X \rightarrowtail Y$ *is* $0 : 0 \to X$.

PROOF.
(i) $0 \to X \to Y$ since $f0 = 0$.
$$\underset{0}{\underset{\big\lfloor \quad 0 \quad \quad f \big\uparrow}{}}$$

(ii) Suppose that $g : W \to X$ with $fg = 0$. Then $fg = f0$; hence, since f is monic, $g = 0$, and we have

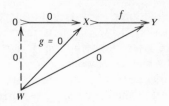

PROPOSITION 4.6*. *Suppose that* $\mathfrak{C}$ *is a category with zero. Then every epic has a cokernel; namely the cokernel of* $f: Y \twoheadrightarrow X$ *is* $0: X \to 0$.

We exemplify these propositions in the category of sets. Since we require a zero, we must consider $\mathfrak{S}_0$, the category of pointed sets and pointed functions. If $f: (X, 0) \to (Y, 0)$, the kernel of f is $f^{-1}(0)$ or, more precisely, the embedding $\kappa: f^{-1}(0) \to X$. If f is monic (thus injective), the kernel is $0 \to X$, that is, it is a singleton set. If $f: (Y, 0) \to (X, 0)$, the cokernel of f is X/fY, or, more precisely, the quotient map $\kappa: X \to X/fY$. By X/fY we mean the pointed set determined by identifying all of fY with the base point while other elements of X retain their identity. If f is epic, that is, surjective, then X/fY is a singleton set.

We should note here a fundamental difference between working within a specific category and working with categories in general. To show that $f: G \to H$ is a monic function in $\mathfrak{S}$, we show that it is injective. To do this we look only *locally*, that is, at f, G, and H. On the other hand, to show that $f: G \to H$ is monic in a category $\mathfrak{C}$, we must search *globally*, that is, look at the entire category. The emphasis here is on what a morphism *does*, rather than what it *is*. The classical definition of an injective function $(x \neq y \Rightarrow f(x) \neq f(y))$ may then be regarded as providing an existence theorem for monomorphisms in $\mathfrak{S}$. Similarly, the classical definitions of injective, surjective, and kernel in $\mathfrak{Ab}$ are existence theorems in $\mathfrak{Ab}$ of the more general categorical notions of monic, epic, and kernel.

The notion of the opposite category enables us to define the important concept of a contravariant functor.

DEFINITION 4.3. A *contravariant functor* $F: \mathfrak{C} \to \mathfrak{D}$ is a functor $F: \mathfrak{C}^{op} \to \mathfrak{D}$. [A functor is sometimes called a *covariant* functor if it is required to emphasize the contrast.] For F to be a contravariant functor, we require explicitly

$$F: \mathfrak{C}(X, Y) \to \mathfrak{D}(FY, FX)$$

and

$$F(fg) = F(g)f(f).$$

We now give some examples.

1. *The contravariant functor* * : $\mathfrak{D}_K \to \mathfrak{D}_K$. Thus V^* is the dual space of

V. If $f: V \rightarrow W$, then $f^*: W^* \rightarrow V^*$ is defined as follows: If $\alpha \in W^*$, $f^*\alpha : V \rightarrow K$ is given by $(f^*\alpha)v = \alpha(fv)$, all $v \in V$. It is easy to see that * is a functor. The contravariant functor * from *finite-dimensional* vector spaces over K to *finite-dimensional* vector spaces over K sets up an isomorphism between the category of finite-dimensional vector spaces over K and its opposite.

2. *The contravariant power set functor* $Q : \mathfrak{S} \rightarrow \mathfrak{S}$. Thus $QX = 2^X$; and if $f : X \rightarrow Y$, $(Qf)B = f^{-1}(B), B \subseteq Y$; thus $Qf : 2^Y \rightarrow 2^X$. Q has the following excellent properties:

$$Qf(B_1 \cup B_2) = Qf(B_1) \cup Qf(B_2),$$

$$Qf(B_1 \cap B_2) = Qf(B_1) \cap Qf(B_2),$$

$$Qf(\varnothing) = \varnothing,$$

$$Qf(Y) = X,$$

$$[Qf(B)]' = Qf(B'),$$

where $'$ stands for complement. These important properties (not possessed by the covariant power set functor P) allow us to consider Q as a functor from $\mathfrak{S}$ to the category of Boolean algebras. Note that to prescribe a topology in X is just to pick a family of elements of 2^X satisfying certain conditions. Then $f : X \rightarrow Y$ is continuous if and only if Qf sends members of the privileged family of 2^Y to members of the privileged family of 2^X. Members of the privileged family are called *open* sets.

3. For a final example we consider $A^- : \mathfrak{S} \rightarrow \mathfrak{S}$. Here $A \in \mathfrak{S}$, $A \neq \varnothing$. On objects, $A^-(B) = A^B$. If $f : B_1 \rightarrow B_2, A^f : A^{B_2} \rightarrow A^{B_1}$ is defined by $A^f(g) = gf$. We have the following diagram:

$$
\begin{array}{ccc}
B_1 & \longmapsto & A^{B_1} \\
{\scriptstyle f}\big\downarrow & & \big\downarrow{\scriptstyle A^f} \\
B_2 & \longmapsto & A^{B_2}
\end{array}
$$

We close this section by making two (related) remarks about procedures for exploiting the duality principle in category theory. Suppose that we take a true statement Σ in some category $\mathfrak{A}$ (for example, $\mathfrak{A} = \mathfrak{G}$, so that we have a theorem of group theory.) We might generalize Σ, that is, find a general categorical statement S such that $S(\mathfrak{A}) = \Sigma$. Now statement S might not be true in all categories, but we might be able to formulate a list of axioms $A_1, A_2, \ldots, A_n$ and then be able to prove the statement $S(\mathfrak{S})$ in categories $\mathfrak{C}$ satisfying these axioms. (Presumably, $\mathfrak{A}$ would be such a category.) Then, for any category $\mathfrak{D}$ satisfying axioms $A_1^{\mathrm{op}},\ldots,A_n^{\mathrm{op}}$, the statement $S^{\mathrm{op}}(\mathfrak{D})$ would be provable. If the axioms are self-dual, or if we can pair them off as dual axioms, we have an especially nice situation. Then, if $S(\mathfrak{C})$ is provable, so is $S^{\mathrm{op}}(\mathfrak{C})$.

Now suppose merely that we can find a statement S that is meaningful in all categories $\mathfrak{C}$ satisfying the self-dual axioms $A_1,\ldots,A_n$ and such that $S(\mathfrak{C})$ specializes to the statement Σ in the category $\mathfrak{A}$. If we could prove $S(\mathfrak{C})$, we would have a proof of $S^{\mathrm{op}}(\mathfrak{C})$, and we could, by specialization, infer a new statement in $\mathfrak{A}$, denoted by Σ'. It is important to realize, however, that we cannot immediately infer the truth of Σ' merely from a proof of Σ. This is because a proof of Σ *in* $\mathfrak{A}$ might have used a property that was not reflected in the axiom system for $\mathfrak{C}$.

5. PRODUCTS AND COPRODUCTS

We have an intuitive feeling that Cartesian products in $\mathfrak{S}$ and direct products in $\mathfrak{G}$ are somehow instances of the same basic concept. We thus seek to generalize to arbitrary categories these ideas. Again, since we do not have elements of sets in general categories we need a characterization in terms of morphisms. The generalization is based on Theorem 5.1 of Chapter 1 (and its analogy in set theory).

DEFINITION 5.1.   A *product* of A_1 and A_2 (if it exists) in the category $\mathfrak{C}$ is an ordered triple $(A;p_1,p_2)$ consisting of an object A and morphisms $p_i:A\to A_i, i=1,2$, as given in diagram (4.1). The notation indicates that, given p_1,p_2,f_1,f_2, a unique morphism f is given by the equation $p_1 f = f_1, p_2 f = f_2$. Note that we have carefully written "a product," since we have not yet proved that a product, if it exists, is unique. This we now do.

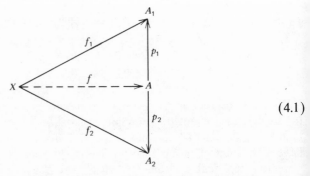

$$(4.1)$$

Suppose that $(A;p_1,p_2)$ is a product and $u:A'\to A$ is a unit. It is then easy to see that $(A';p_1u,p_2u)$ is also a product. In fact, this is all the choice we have in finding products. We say that the products $(A;p_1,p_2)$, $(A';p_1u,p_2u)$ are *canonically equivalent*.

THEOREM 5.1. *A product, if it exists, is unique, up to canonical equivalence.*

PROOF. The proof is essentially that of Theorem 5.1 of Chapter 1. Suppose that both $(A;p_1,p_2)$ and $(A';p_1',p_2')$ are products. We have two diagrams:

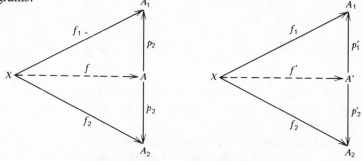

Since A is a product there exists a unique $u:A'\to A$ such that $p_1u=p_1'$ and $p_2u=p_2'$.

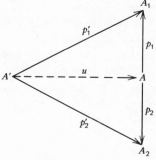

Of course, we have still to show that u is a unit. Now, since A' is a product there exists a unique $v:A\to A'$ such that $p_1'v=p_1$ and $p_2'v=p_2$.

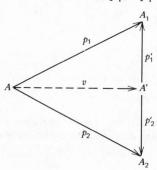

Thus $p_1uv = p_1 = p_1 1$, and $p_2uv = p_2 = p_2 1$.

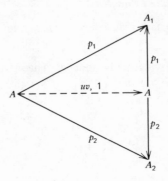

Since A is a product we conclude from the *uniqueness* part of the definition that $uv = 1$. Similarly we deduce that $vu = 1$. Thus $u : A' \to A$ is a unit such that $p_1u = p_1'$ and $p_2u = p_2'$, as claimed.

The reader can readily see that the degree of uniqueness proved is all one would expect—and all one needs. It means that if two candidates for the product of A_1 and A_2 emerge, there is a perfectly definite unit transforming one to the other and these transformations are consistent in a quite precise and highly important sense.

In $\mathfrak{S}$ the product is the Cartesian product.[†] In $\mathfrak{G}$ it is the direct product. In $\mathfrak{Ab}$ it is the direct sum. Let us consider the category $\mathfrak{C}_S$ where S is a preordered set. Recall that $f : A \to B$ in $\mathfrak{C}_S$ simply tells us that $A \leqslant B$. Let $(A; p_1, p_2)$ be the product of A_1 and A_2 in $\mathfrak{C}_S$.

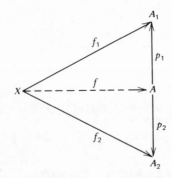

The product diagram can now be interpreted as saying: $A \leqslant A_1, A \leqslant A_2$ and if $X \leqslant A_1, X \leqslant A_2$ then $X \leqslant A$. In other words, the product A of A_1 and A_2 is

---

[†]We should say "…the Cartesian product together with the two projections," but we permit ourselves this informal terminology once the point has been established.

the greatest lower bound of A_1 and A_2. The statement that the product (g.l.b.) is essentially unique means that any other greatest lower bound A' is related to A by a unit, that is, $A' \leqslant A$ and $A \leqslant A'$. We see, therefore, that in an ordered set the g.l.b. is strictly unique.

We have thus generalized the notion of Cartesian product to arbitrary categories; that is, we have provided a definition in an arbitrary category $\mathfrak{C}$ which specializes to the Cartesian product in $\mathfrak{S}$. Our method of definition of the product is called *definition by universal property*, and is characteristic of categorical formulations. We have already seen examples of this type of definition in connection with *kernels* and *cokernels*. Again we notice that the Cartesian product description serves as an existence theorem for products in the category $\mathfrak{S}$; and that, whereas one defines the Cartesian product of A_1 and A_2 merely in terms of the sets A_1, A_2 themselves, the product of two objects of a category makes reference to the totality of objects of the category. We may say that traditional definitions are *local* in character, while definitions by universal properties are *global*. We would also like to insist on the vivid phrase that traditional definitions describe what a mathematical construct *is*, whereas universal, categorical definitions describe what it *does*, what role it plays in the category.

The next examples demonstrate that the product of objects depends very much on what category we are considering. In $\mathfrak{G}$, the product of $\mathbf{Z}_2$ and $\mathbf{Z}_4$ is the direct product, $\mathbf{Z}_2 \times \mathbf{Z}_4$. In $\mathfrak{Ab}$, it is the direct sum, $\mathbf{Z}_2 \oplus \mathbf{Z}_4$ (really the same thing). In the category of cyclic groups the product of $\mathbf{Z}_2$ and $\mathbf{Z}_4$ does not exist. Notice that this asserts more than that $\mathbf{Z}_2 \oplus \mathbf{Z}_4$ is not cyclic. We must show that *no* cyclic group serves as product. Suppose then that the product of $\mathbf{Z}_2$ and $\mathbf{Z}_4$ is $\mathbf{Z}_m$. We have

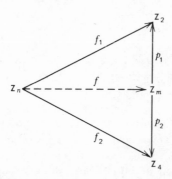

Since both f_1 and f_2 could be surjective functions, p_1 and p_2 must be surjective functions. In particular, $\mathbf{Z}_m \twoheadrightarrow \mathbf{Z}_4$ means that $4|m$. If $\mathbf{Z}_m$ is a product, we have the following diagram.

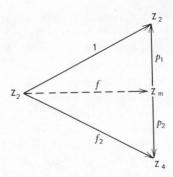

This shows that Z_2 is a direct summand of Z_m. Since $4|m$, we have arrived at a contradiction.

From now on we will denote the product of A_1 and A_2 by $A_1 \times A_2$ and the morphism f in the product diagram by $\{f_1, f_2\}$.

The definition of the product of two objects can, easily and obviously, be extended to the product of any number of objects. In the case of a finite number of objects we denote the projections by $p_1, \ldots, p_n$ and we write $f = \{f_1, \ldots, f_n\}$. For the infinite case we write $f = \{f_i\}_{i \in I}$. We call the morphism $f_i = p_i f$ the ith *component* of f.

Suppose that $g_i : X_i \to Y_i$, $i = 1, 2$. We can define a morphism $g_1 \times g_2 :$ $X_1 \times X_2 \to Y_1 \times Y_2$ by $g_1 \times g_2 = \{g_1 p_1, g_2 p_2\}$. This accords with our usual notation in $\mathfrak{S}$ and $\mathfrak{G}$.

PROPOSITION 5.2.

  (i) $(g_1 \times g_2)(h_1 \times h_2) = (g_1 h_1 \times g_2 h_2)$. *Thus* $\underline{\quad} \times \underline{\quad}$ *is a functor from* $\mathfrak{C} \times \mathfrak{C}$ *to* $\mathfrak{C}$.

  (ii) $(g_1 \times g_2) o \{f_1, f_2\} = \{g_1 f_1, g_2 f_2\}$.

  (iii) $\{f_1, f_2\} g = \{f_1 g, f_2 g\}$.

PROOF OF (ii). We have $A \overset{\{f_1, f_2\}}{\to} X_1 \times X_2 \overset{g_1 \times g_2}{\to} Y_1 \times Y_2$. We must check the components. Now $p_i(g_1 \times g_2)\{f_1, f_2\} = g_i p_i \{f_1, f_2\} = g_i f_i$, $i = 1, 2$. The reader should prove assertions (i) and (iii).

PROPOSITION 5.3. *If any two objects of* $\mathfrak{C}$ *have a product, then so does any finite number of objects. In fact,* $(X_1 \times X_2) \times X_3$ *is the product of* X_1, X_2 *and* X_3.

The proof uses induction, and it is clearly sufficient to establish the final assertion. To show that $(X_1 \times X_2) \times X_3$ is the product of X_1, X_2, and X_3 consider the following projections:

$$(X_1 \times X_2) \times X_3 \overset{p_1 o p_1}{\to} X_1,$$

$$(X_1 \times X_2) \times X_3 \overset{p_1 op_2}{\to} X_2,$$

$$(X_1 \times X_2) \times X_3 \overset{p_2}{\to} X_3.$$

(Notice that we are using the symbols p_1, p_2 generically. For example, in $p_1 op_1$, the second p_1 is $(X_1 \times X_2) \times X_3 \to X_1 \times X_2$ and the first p_1 is $X_1 \times X_2 \to X_1$.) We leave it to the reader to verify that the universal property for a product is now satisfied.

EXAMPLES.

1. *The category of finitely generated abelian groups.* Any two objects have a product: the direct sum. No infinite set of nonzero objects has a product. To show this, it is not enough to demonstrate that the direct sum of an infinite number of nonzero objects in the category is not finitely generated. We must show that no finitely generated abelian group will work. That this is really necessary is illustrated by the next example.

2. *The category of torsion abelian groups.* For a finite number of objects the product is the direct sum. Suppose that we take a fixed prime p and consider the collection $\{Z_{p^n}\}$ of cyclic groups of order $p^n, n = 1, 2, \ldots$. Then ΠZ_{p^n} is not torsion, because $\{1_n\}$, the elements whose nth component is the generator of the nth group, has infinite order. Thus ΠZ_{p^n} is not the product in this category. Consider, however, $T(\Pi Z_{p^n})$, the torsion subgroup of ΠZ_{p^n}. In general, if we have $(A_i)_{i \in I}$ a collection of torsion abelian groups, $T(\Pi A_i)$ is the product in the category of torsion abelian groups. This is because in $\mathfrak{G}$ we have

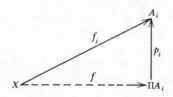

But if X is a torsion group, so is fX, and thus $fX \subseteq T(\Pi A_i)$. We therefore have

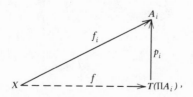

by restricting the domain of p_i and the codomain of f above.

   3. *The category of finite abelian groups.* Again finite products exist, but in this category products of an infinite number of nonzero objects generally do not exist.

DEFINITION 5.2. $F : \mathfrak{C} \to \mathfrak{D}$ is a *product-preserving* functor if $(F(X_1 \times X_2); Fp_1, Fp_2)$ is the product of FX_1 and FX_2 in $\mathfrak{D}$.

   Given objects X_1 and X_2 in $\mathfrak{C}$ we have, for any functor $F : \mathfrak{C} \to \mathfrak{D}$, a diagram in $\mathfrak{D}$:

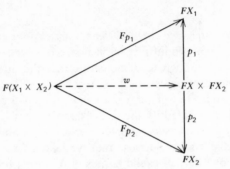

To see if F is product-preserving amounts to checking that w is a unit. If so, we write $Fp_1 = p_1, Fp_2 = p_2$. Also, we plainly then have

$$F\{f_1, f_2\} = \{Ff_1, Ff_2\}.$$

EXAMPLES.
   1. Underlying functors are product-preserving.
   2. Abel is product-preserving.
   3. The free functor from $\mathfrak{S}$ to $\mathfrak{G}$ is not product-preserving.
We will next dualize the notion of product.

DEFINITION 5.1*.   The *coproduct* of objects A_1 and A_2 in $\mathfrak{C}$ is an ordered triple $(A; q_1, q_2)$ such that

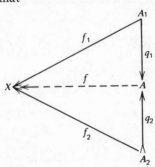

We immediately know that the coproduct is unique up to canonical equivalence.

In $\mathfrak{S}$ and $\mathfrak{T}$ the coproduct is the disjoint union. In $\mathfrak{S}_0, \mathfrak{T}_0$ (sets with base point, spaces with base point), it is the wedge union, that is, the disjoint union with the base points identified. In $\mathfrak{G}$ the coproduct of two groups is the *free product*. If G_1 and G_2 are given by generators and relators, the free product is given by the (disjoint) unions of the generators and the relators. Recall that if G is given by generators and relators, then to give a homomorphism $f: G \to X$ is to give a function from the generators to X, which, when extended to a homomorphism from the free group on the generators to X, takes the relators of G to the neutral element of X. This observation enables us to conclude immediately that the free product is the coproduct in $\mathfrak{G}$.

In $\mathfrak{Ab}$ the coproduct is the direct sum. If S is a preordered set, the coproduct in $\mathfrak{C}_S$ is the least upper bound.

Let $\mathfrak{C}$ be the category determined by the set of integers preordered by divisibility ($x \leqslant y$ if and only if $x|y$). Then the product, or g.l.b., is the greatest common divisor, and the coproduct, or l.u.b., is the least common multiple. We can therefore immediately say:

$$gcd(gcd(x,y),z) = gcd(x, gcd(y,z)),$$

$$lcm(lcm(x,y),z) = lcm(x, lcm(y,z)).$$

Then *gcd* and *lcm* are unique up to units, which are, in this case, just ± 1. These statements are among the trivial parts of the theory of divisibility. They are universal facts. A nonuniversal fact that is not proved on the categorical level is the following: *in a principal ideal domain, gcd(a,b) is a linear combination of a and b.*

The following notations will be used systematically— and in generalized form—for products and coproducts:

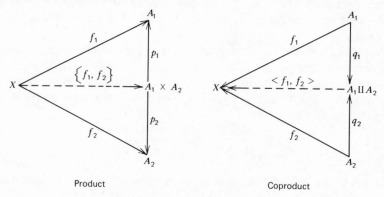

Product   Coproduct

Suppose that in $\mathfrak{C}$, a category with finite products, coproducts, and zero, we have $X_1,\ldots,X_m; Y_1,\ldots,Y_n$. Let $f_{ij}:X_i\to Y_j$ for $1\leqslant i\leqslant m, 1\leqslant j\leqslant n$, in $\mathfrak{C}$. We can now define a morphism

$$X_1\amalg X_2\amalg\cdots\amalg X_m\to Y_1\times Y_2\times\cdots\times Y_n.$$

There are apparently two ways to do this. We could describe how the morphism takes $\amalg X_i$ into $Y_1, Y_2,\ldots,Y_n$. We get a morphism

$$f=\{\langle f_{11},f_{21},\ldots,f_{m1}\rangle,\langle f_{12},f_{22},\ldots,f_{m2}\rangle,\ldots,\langle f_{1n},f_{2n},\ldots,f_{mn}\rangle\}.$$

Or we could describe how X_1 goes into $\Pi Y_i, X_2$ into $\Pi Y_i,\ldots,X_m$ into ΠY_i. We get a morphism

$$f'=\langle\{f_{11},f_{12},\ldots,f_{1n}\},\{f_{21},f_{22},\ldots,f_{2n}\},\ldots,\{f_{m1},f_{m2},\ldots,f_{mn}\}\rangle.$$

THEOREM 5.4. *The two morphisms* f, f' *described above are equal.*

PROOF. We must show the components of each are the same. The components, with respect to the product structure in the codomain, of f being obvious, we proceed to compare those of f'. We have

$$p_j f'=p_j\langle\ldots,\{f_{k1},f_{k2},\ldots,f_{kn}\},\ldots\rangle$$

$$=\langle\ldots,p_j\{f_{k1},f_{k2},\ldots,f_{kn}\},\ldots\rangle \qquad [\text{dual of Proposition 5.2(iii)}]$$

$$=\langle f_{1j},f_{2j},\ldots,f_{mj}\rangle$$

$$=p_j f.$$

This result can be extended to infinite products. It justifies the use of a matrix notation $f=(f_{ij})$ for the morphism described above! Of course, we have so far made no use of the assumption that $\mathfrak{C}$ has a zero object, but we do so now.

We warned in Section 4 that one cannot pass with certainty from a theorem in a specific category to its "dual." We now give an example where the dual of a true theorem is, in fact, false. Let $m=n$ (possibly infinite) and $X_i=Y_i$ and let $f_{ij}=1$ if $i=j, f_{ij}=0$ otherwise. We get the *coproduct-product morphism*

$$\omega:X_1\amalg\ldots\amalg X_m\to X_1\times\cdots\times X_m.$$

(The reader should have no trouble in representing ω as a natural transformation of functors.) An example of ω is the canonical homomorphism in $\mathfrak{G}$ of the free product to the direct product. Now take the case of a countable collection $\{A_i\}$ of abelian groups. The following statement is true in $\mathfrak{Ab}$ and is expressed in categorical language:

$$\oplus A_i \xrightarrow{\omega} \Pi A_i \text{ is monic.}$$

The dual statement is false, however. It is easy to see that ω is self-dual, so that the dual statement is:

$$\Pi A_i \xleftarrow{\omega} \oplus A_i \text{ is epic.}$$

Note here that ΠA_i consists of sequences $\{a_i\}, a_i \in A_i$; while $\oplus A_i$ consists of sequences $\{a_i\}, a_i \in A_i$, all but finitely many a_i being equal to zero. Thus the true statement in $\mathfrak{Ab}$, that ω is monic, cannot be true in all abelian categories (see Section 8), since the axioms of an abelian category are self-dual.

Of course, dual to Definition 5.2 we have the concept of a *coproduct-preserving* functor. Examples will be found in the Exercises.

6. PULLBACKS AND PUSHOUTS

We first consider several examples. Suppose that A is a set and $A_1, A_2 \subseteq A$. The set $A_1 \cap A_2$ gives the commutative diagram

$$
\begin{array}{ccc}
A_1 \cap A_2 & \cdots\!\!\!\rightarrow & A_1 \\
\vdots & & \downarrow \\
A_2 & \longrightarrow & A
\end{array}
$$

where all the maps are inclusion maps. In $\mathfrak{G}$, $\mathfrak{Ab}$, and $\mathfrak{T}$ we have the same situation. In all these cases we have an intuitive feeling that we have obtained a commutative square in the best possible way.

As another example, suppose that we have, in $\mathfrak{G}$, a diagram of groups and homomorphisms

$$
\begin{array}{ccc}
& & G_1 \\
& & \downarrow{\phi_1} \\
G_2 & \xrightarrow[\phi_2]{} & H
\end{array}
$$

Then if G_0 is the subgroup of $G_1 \times G_2$ consisting of all pairs (g_1, g_2) such that $\phi_1 g_1 = \phi_2 g_2$, we have a commutative diagram

$$
\begin{array}{ccc}
G_0 & \overset{\psi_1}{\dashrightarrow} & G_1 \\
{\scriptstyle \psi_2}\downarrow & & \downarrow{\scriptstyle \phi_1} \\
G_2 & \underset{\phi_2}{\longrightarrow} & H
\end{array}
$$

in which ψ_1, ψ_2 are the restrictions to G_0 of the projections of $G_1 \times G_2$ onto G_1, G_2 respectively. Again this seems to be an example where we have picked the best completion of the diagram to a commutative square. We will shortly be giving precise objective meaning to these intuitions.

As an example of the dual situation, suppose that A_1, A_2 are abelian groups and that B is a subgroup of each. Thus we have a diagram of inclusions

$$
\begin{array}{ccc}
B & \longrightarrow & A_1 \\
\downarrow & & \\
A_2 & &
\end{array}
$$

Now if we form the diagram

$$
\begin{array}{ccc}
B & \longrightarrow & A_1 \\
\downarrow & & \downarrow \\
A_2 & \longrightarrow & A_1 \oplus A_2
\end{array}
$$

where the arrows $A_i \to A_1 \oplus A_2$ are the canonical embeddings into the direct sum, $i = 1, 2$, then the diagram fails to commute. We may, however, force it to commute by factoring out of $A_1 \oplus A_2$ the subgroup consisting of elements $(b, -b), b \in B$. Write A for this quotient group of $A_1 \oplus A_2$ and let $A_i \to A, i = 1, 2$, be the composite of the canonical embedding $A_i \to A_1 \oplus A_2$ and the projection $A_1 \oplus A_2 \to A$. Now the diagram

$$
\begin{array}{ccc}
B & \longrightarrow & A_1 \\
\downarrow & & \downarrow \\
A_2 & \longrightarrow & A
\end{array}
$$

commutes, and we again have the feeling that we have extended our original diagram (consisting of the inclusions $B \subseteq A_i, i = 1, 2$) to a commutative square in a very economical way. We now move toward precise definitions of constructions of which the examples cited will turn out to be special cases.

Going back to the first example, we would like to characterize $A_1 \cap A_2$ in terms of morphisms. Notice that any function $X \to A$ with image in both A_1 and A_2 actually has its image in $A_1 \cap A_2$. Thus we have

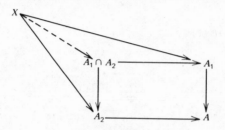

All of the other examples discussed yield diagrams like this one.

DEFINITION 6.1. A *pullback* of

in $\mathfrak{C}$ is a commutative square

such that the following universal property holds:

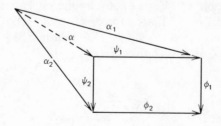

Recall that, by our notational conventions, the diagram implies that $\phi_1 \alpha_1 = \phi_2 \alpha_2$ and that there then exists a *unique* morphism α such that $\psi_1 \alpha = \alpha_1, \psi_2 \alpha = \alpha_2$.

DEFINITION 6.1*.   A *pushout* of

in $\mathfrak{C}$ is a commutative square

such that the following universal property holds:

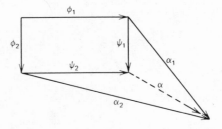

The reader should now check that the examples given at the beginning of this section are all pullbacks or pushouts.

THEOREM 6.1.   *The pullback is unique up to canonical unit.*

The proof is similar to the proof of uniqueness of the product, and we omit it. There is, in fact, a close relation between pullback and product. Let Y be a fixed object of $\mathfrak{C}$, and consider the category $\mathfrak{C}/Y$ of $\mathfrak{C}$-*objects over* Y. The objects of $\mathfrak{C}/Y$ are morphisms $\phi:X\to Y$ of $\mathfrak{C}$. If $\phi:X\to Y$ and $\psi:Z\to Y$ are objects of $\mathfrak{C}/Y$, a morphism $f:\phi\to\psi$ of $\mathfrak{C}/Y$ is a morphism $f:X\to Z$ of $\mathfrak{C}$, such that $\psi f=\phi$,

Essentially, a pullback of

$$\downarrow \phi_1$$

$$\xrightarrow[\phi_2]{}$$

in $\mathfrak{C}$ is a product of ϕ_1 and ϕ_2 in $\mathfrak{C}/Y$. For suppose that we have a pullback

$$
\begin{array}{ccc}
X_0 & \xrightarrow{\psi_1} & X_1 \\
\psi_2 \downarrow & & \downarrow \phi_1 \\
X_2 & \xrightarrow[\phi_2]{} & Y
\end{array}
$$

Then we have

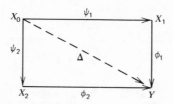

and $(\Delta; \psi_1, \psi_2)$ is the product of ϕ_1 and ϕ_2 in $\mathfrak{C}/Y$, as the reader will readily check. Also if $\mathfrak{C}$ has a zero object 0, then the pullback of $X_1 \to 0, X_2 \to 0$ is the pair of projections $p_i : X_1 \times X_2 \to X_i$, $i = 1, 2$.

We now show, by an example, how we may reason about pullbacks in an arbitrary category.

THEOREM 6.2. *Suppose that $\mathfrak{C}$ is a category with zero and let*

$$
\begin{array}{ccc}
X_0 & \xrightarrow{\psi_1} & X_1 \\
\psi_2 \downarrow & & \downarrow \phi_1 \\
X_2 & \xrightarrow[\phi_2]{} & Y
\end{array}
$$

be a pullback diagram in $\mathfrak{C}$. Then
  (i) If ψ_2 has kernel κ, then ϕ_1 has kernel $\psi_1 \kappa$,
  (ii) If ϕ_1 has kernel λ, then $\lambda = \psi_1 \kappa$ for some kernel κ of ψ_2.

PROOF.   To prove (i) we have the following diagram. We know that κ is a kernel and must show that $\psi_1\kappa$ is also.

$$
\begin{array}{ccc}
K & = & K \\
{\scriptstyle\kappa}\downarrow & {\scriptstyle\psi_1} & \downarrow{\scriptstyle\psi_1\kappa} \\
X_0 & \xrightarrow{} & X_1 \\
{\scriptstyle\psi_2}\downarrow & & \downarrow{\scriptstyle\phi_1} \\
X_2 & \xrightarrow[\phi_2]{} & Y
\end{array}
$$

First, $\phi_1\psi_1\kappa = \phi_2\psi_2\kappa = 0$ as $\psi_2\kappa = 0$. Next we must show that $\psi_1\kappa$ is monic. To do this, we note that the uniqueness part of the pullback definition says that $\psi_1\alpha = \psi_1\beta$ and $\psi_2\alpha = \psi_2\beta$ together imply $\alpha = \beta$. To prove $\psi_1\kappa$ monic, suppose that $\psi_1\kappa\rho = \psi_1\kappa\sigma$. We also have $\psi_2\kappa\rho = \psi_2\kappa\sigma(=0)$. Thus $\kappa\rho = \kappa\sigma$ and so, since κ is monic, $\rho = \sigma$. Thus $\psi_1\kappa$ is monic. We still must find β in the diagram below.

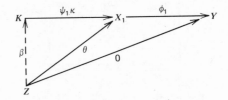

Then $\phi_1\theta = 0 = \phi_2 0$, so that, by the pullback property, there exists $\alpha : Z \to X_0$ with $\psi_1\alpha = \theta, \psi_2\alpha = 0$. Since κ is the kernel of $\psi_2, \alpha = \kappa\beta$ for some $\beta : Z \to K$, and then $\theta = \psi_1\alpha = \psi_1\kappa\beta$, as required.

To prove (ii) we suppose that ϕ_1 has kernel λ.

$$
\begin{array}{ccc}
K & & K \\
{\scriptstyle\kappa}\vdots\;\downarrow & & \downarrow{\scriptstyle\lambda} \\
X_0 & \xrightarrow{\psi_1} & X_1 \\
{\scriptstyle\psi_2}\downarrow & & \downarrow{\scriptstyle\phi_1} \\
X_2 & \xrightarrow[\phi_2]{} & Y
\end{array}
$$

By the same reasoning that gave us α above, we can find $\kappa : K \to X_0$ such that $\psi_1\kappa = \lambda$ and $\psi_2\kappa = 0$. Moreover κ is monic as λ is monic. We still must show that κ is the kernel of ψ_2. That is, we must find σ to complete the diagram

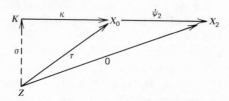

Now $\phi_1\psi_1\tau=\phi_2\psi_2\tau=0$, so that there exists σ with $\psi_1\tau=\lambda\sigma$. In fact, this morphism σ is what we are looking for. We must show that $\kappa\sigma=\tau$. But we know that $\psi_1\kappa\sigma=\lambda\sigma=\psi_1\tau$ and $\psi_2\kappa\sigma=0=\psi_2\tau$. Thus $\kappa\sigma=\tau$ and the proof is complete.

This theorem may be paraphrased as saying that the kernel of ϕ_1 exists if and only if the kernel of ψ_2 exists, and the kernel objects are then the same. Of course, in many applications we already know that kernels exist throughout the category. This proof is then unduly fussy—the first half of the argument would suffice.

As an application, consider the pullback square in $\mathfrak{G}$:

$$
\begin{array}{ccc}
G_0 & \xrightarrow{\psi_1} & G_1 \\
{\scriptstyle\psi_2}\downarrow & & \downarrow{\scriptstyle\phi_1} \\
G_2 & \xrightarrow[\phi_2]{} & G
\end{array}
$$

The theorem tells us the kernel objects of ψ_2 and ϕ_1 are the same.

$$
\begin{array}{ccc}
K & = & K \\
\downarrow & & \downarrow \\
G_0 & \xrightarrow{\psi_1} & G_1 \\
{\scriptstyle\psi_2}\downarrow & & \downarrow{\scriptstyle\phi_1} \\
G_2 & \xrightarrow[\phi_2]{} & G
\end{array}
$$

In fact, we note that, since G_0 is the subgroup of $G_1 \times G_2$ consisting of pairs (g_1,g_2) with $\phi_1 g_1 = \phi_2 g_2$, then $\ker\psi_2$ consists of pairs (g_1,e) with $\phi_1 g_1 = e$ and so $(g_1,e) \mapsto g_1$ yields an isomorphism from $\ker\psi_2$ to $\ker\phi_1$.

As an application of the dual theorem, consider the pushout square in $\mathfrak{Ab}$, discussed earlier:

$$
\begin{array}{ccc}
B & \rightarrowtail & A_1 \\
\downarrow & & \downarrow{\scriptstyle\psi_1} \\
A_2 & \xrightarrow[\psi_2]{} & A
\end{array}
$$

Here A is the quotient of $A_1 \oplus A_2$ by the subgroup B_0, consisting of pairs $(b, -b)$, $b \in B$. We note that ψ_1 (and ψ_2) is monic. For $\psi_1(a_1) = (a_1, 0)$ $\bmod B_0$, so that $\psi_1(a_1) = 0$ implies that $(a_1, 0) = (b, -b)$ for some $b \in B$. But then $b = 0, a_1 = 0$. Our theorem, in its dual form, claims that the cokernel of ψ_1 is isomorphic to A_2/B. We map A to A_2/B by $(a_1, a_2) \bmod B_0 \mapsto a_2$ $\bmod B$. This is obviously a surjection, and its kernel consists precisely of the elements $(a_1, 0) \bmod B_0$, that is, its kernel is the image of ψ_1.

Suppose now that we have a pullback

$$
\begin{array}{ccc}
A & \overset{\psi_1}{\longrightarrow} & X_1 \\
{\scriptstyle \psi_2}\downarrow & & \downarrow{\scriptstyle \phi_1} \\
X_2 & \underset{\phi_2}{\longrightarrow} & Y
\end{array}
$$

in $\mathfrak{Ab}$. We can think of ψ_1, ψ_2 as a single morphism into the product, which is just the direct sum:

$$ A \overset{\{\psi_1, \psi_2\}}{\to} X_1 \oplus X_2 $$

The uniqueness part of the pullback property is equivalent to the assertion that $\{\psi_1, \psi_2\}$ is monic; for $\{\psi_1, \psi_2\}u = \{\psi_1, \psi_2\}v$ means that $\psi_1 u = \psi_1 v, \psi_2 u = \psi_2 v$. Writing ϕ_1, ϕ_2 as one morphism with domain the coproduct, which is again the direct sum, we obtain the diagram

$$ A \overset{\{\psi_1, \psi_2\}}{\rightarrowtail} X_1 \oplus X_2 \overset{\langle \phi_1 \phi_2 \rangle}{\to} Y. $$

The composite is the morphism $\phi_1 \psi_1 + \phi_2 \psi_2$, as is easily checked. We would like the commutativity of the pullback square to be reflected in the fact that the composite morphism is 0. To achieve this we change the sign of one morphism, say ψ_2, and obtain

$$ A \overset{\{\psi_1, -\psi_2\}}{\rightarrowtail} X_1 \oplus X_2 \overset{\langle \phi_1, \phi_2 \rangle}{\to} Y $$

Now the composite is $\phi_1 \psi_1 - \phi_2 \psi_2 = 0$, and of course, $\{\psi_1, -\psi_2\}$ is still monic. We get the following results.

PROPOSITION 6.3 *The diagram*

$$A \xrightarrow{\psi_1} X_1$$
$$\psi_2 \downarrow \qquad \downarrow \phi_1$$
$$X_2 \xrightarrow[\phi_2]{} Y$$

commutes if and only if

$$A \xrightarrow{\{\psi_1, -\psi_2\}} X_1 \oplus X_2 \xrightarrow{\langle \phi_1, \phi_2 \rangle} Y$$
$$\underset{0}{\overline{}} \uparrow$$

PROPOSITION 6.4. *The diagram*

$$A \xrightarrow{\psi_1} X_1$$
$$\psi_2 \downarrow \qquad \downarrow \phi_1$$
$$X_2 \xrightarrow[\phi_2]{} Y$$

is a pullback if and only if $\{\psi_1, -\psi_2\}$ *is the kernel of* $\langle \phi_1, \phi_2 \rangle$.

PROOF. Compare the two diagrams below. The pullback diagram on the left implies the kernel diagram on the right, and conversely.

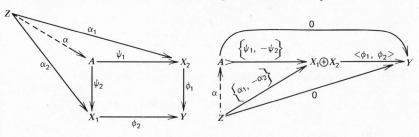

PROPOSITION 6.4*. *The diagram*

$$\xrightarrow{\psi_1}$$
$$\psi_2 \downarrow \qquad \downarrow \phi_1$$
$$\xrightarrow[\phi_2]{}$$

is a pushout if and only if $\langle \phi_1, \phi_2 \rangle$ *is the cokernel of* $\{\psi_1, -\psi_2\}$.

7. ADJOINT FUNCTORS

Adjoint functors were first introduced by D. M. Kan. They provide a generalization of the exponential law of set theory:

$$\mathfrak{S}(A \times B, C) \cong \mathfrak{S}(A, C^B), A, B, C \in \mathfrak{S},$$

where C^B is the set of functions from B to C. Consider the functor $F = - \times B$ from $\mathfrak{S}$ to $\mathfrak{S}$ and the functor $G = -^B$ from $\mathfrak{S}$ to $\mathfrak{S}$. That is, $FA = A \times B$ and $GA = A^B$. Then for pairs of objects A, C the exponential law above becomes:

$$\overset{\eta_{AC}}{\mathfrak{S}(FA, C) \cong \mathfrak{S}(A, GC)}. \tag{7.1}$$

Given an ordered pair of objects (A, C) of $\mathfrak{S}$ we get a set $\mathfrak{S}(FA, C)$, and this gives us a functor from $\mathfrak{S} \times \mathfrak{S}$ to $\mathfrak{S}$, contravariant in the first argument and covariant in the second argument. Thus we have a functor $\mathfrak{S}^{\mathrm{op}} \times \mathfrak{S} \to \mathfrak{S}$. Similarly, $(A, C) \mapsto \mathfrak{S}(A, GC)$ is also a functor $\mathfrak{S}^{\mathrm{op}} \times \mathfrak{S} \to \mathfrak{S}$, and (7.1) says that these two functors are naturally equivalent. Let us now express precisely what this naturality means. Suppose that we have $A_1 \overset{\alpha}{\to} A$, $C \overset{\beta}{\to} C_1$. We have (writing η for η_{AC})

$$
\begin{array}{ccccccc}
FA_1 & \overset{F\alpha}{\to} & FA & \overset{\phi}{\to} & C & \overset{\beta}{\to} & C_1 \\
& & \downarrow{\eta} & & & & \\
A_1 & \overset{\alpha}{\to} & A & \overset{\eta(\phi)}{\to} & GC & \overset{G\beta}{\to} & GC_1
\end{array}
$$

The following equation expresses the naturality of the equivalence η:

$$\eta(\beta o \phi o F\alpha) = G\beta o \eta(\phi) o \alpha.$$

The reader should verify this equation.

As a second motivating example, consider $\mathrm{Fr}(S)$, the free abelian group on S as basis. We have the following diagram.

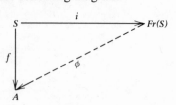

In this diagram i and f are maps of sets, although $\text{Fr}(S)$ and A are abelian groups. The diagram asserts that there exists a unique $\phi \in \mathfrak{Ab}(\text{Fr}(S), A)$ such that $\phi i = f$. The diagram above has arrows in two categories. We can remedy this by use of the underlying functor $U: \mathfrak{Ab} \to \mathfrak{S}$. In $\mathfrak{S}$ we have

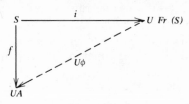

Thus given $f \in \mathfrak{S}(S, UA)$ we find a map $\phi \in \mathfrak{Ab}(\text{Fr}(S), A)$. We have here a relation between the underlying and free functors very much like the earlier relation between $-^B$ and $- \times B$, namely

$$\mathfrak{Ab}(\text{Fr}(S), A) \overset{\eta}{\cong} \mathfrak{S}(S, UA),$$

where $\eta(\phi) = f$. In particular, if $A = \text{Fr}(S)$, the correspondence

$$\mathfrak{Ab}(\text{Fr}(S), \text{Fr}(S)) \cong \mathfrak{S}(S, U\text{Fr}(S))$$

takes $1_{\text{Fr}(S)}$ to $i: S \to U(\text{Fr}(S))$. Thus $i = \eta(1_{\text{Fr}(S)})$. The naturality rule then implies that $U\phi \circ i = f$, as desired. Notice that, in this second example, the functors run to and fro between two categories, $\mathfrak{S}$ and $\mathfrak{Ab}$; this will be the usual situation for adjoint functors. We are led to the following definition.

DEFINITION 7.1. The functor $F: \mathfrak{C} \to \mathfrak{D}$ is said to be *left adjoint* to the functor $G: \mathfrak{D} \to \mathfrak{C}$ if there exists a natural equivalence $\eta = \eta_{AB}$,

$$\eta: \mathfrak{D}(FA, B) \cong \mathfrak{C}(A, GB),$$

where $A \in \mathfrak{C}$, $B \in \mathfrak{D}$. We also say that G is *right adjoint* to F, and that the *adjugant* η establishes an *adjunction* between F and G. We write $F \overset{\eta}{\dashv} G$. We repeat the specific criterion of naturality:

$$\eta(\beta \phi \circ F\alpha) = G\beta \circ \eta \phi \circ \alpha, \tag{7.2}$$

for $\alpha: A_1 \to A$ in $\mathfrak{C}$, $\beta: B \to B_1$ in $\mathfrak{C}$.

Very important special cases of (7.2) are

$$\text{(i) } \eta(\beta \circ \phi) = G\beta \circ \eta \phi, \qquad \text{(ii) } \eta(\phi \circ F\alpha) = \eta \phi \circ \alpha. \tag{7.3}$$

EXAMPLES.

1. Let $\mathfrak{C}_S$ be the category corresponding to the set of real numbers, $\mathbf{R}$, ordered by $\leqslant$, and let $\mathfrak{C}_T$ be the category corresponding to the set of integers, $\mathbf{Z}$, ordered by $\leqslant$. Recall that functors between such categories are just order-preserving functions. Let $\mathrm{Emb}:\mathfrak{C}_T \to \mathfrak{C}_S$ be the embedding functor and $N:\mathfrak{C}_S \to \mathfrak{C}_T$ be the functor $Nx =$ next highest integer to x. Then $N \dashv \mathrm{Emb}$, for $\mathfrak{C}_T(Nx,m) \cong \mathfrak{C}_S(x, \mathrm{Emb}\, m)$, where $x \in \mathbf{R}$ and $m \in \mathbf{Z}$. The reader should examine the (simple) meaning of this abstract statement.

2. The free and underlying functors in various categories give examples of pairs of adjoint functors.

3. Abel $\dashv U$ where Abel: $\mathfrak{G} \to \mathfrak{Ab}$ and U is the underlying functor $U:\mathfrak{Ab} \to \mathfrak{G}$. That is, remembering Abel $(G) = G/G'$,

$$\mathfrak{Ab}(G/G', A) \cong \mathfrak{G}(G, UA).$$

The reader should check this equivalence—it is, in fact, even an isomorphism of abelian groups.

4. Given a group G, written multiplicatively, we can form the integral group ring, $\mathbf{Z}(G)$. This is the ring whose additive structure is the free abelian group on the set G and whose multiplication is given by the group multiplication of G. [Since G (additively) generates $\mathbf{Z}(G)$ it is sufficient to give the multiplication on G.] Then $G \mapsto \mathbf{Z}(G)$ gives a functor $\mathfrak{G} \to \mathfrak{R}_1$, where $\mathfrak{R}_1$ is the category of rings with unity element (see Section 1 of Chapter 4). Given a ring R, denote by $\mathrm{Un}(R)$ the group of units of R. This gives a functor $\mathfrak{R}_1 \to \mathfrak{G}$. We have a pair of adjoint functors, that is,

$$\mathfrak{R}_1(\mathbf{Z}(G), R) \overset{\eta}{\cong} \mathfrak{G}(G, \mathrm{Un}\, R).$$

To see this, let $f \in \mathfrak{R}_1(\mathbf{Z}(G), R)$. Since f preserves the unity element, it preserves the units. But elements of G are units of $\mathbf{Z}(G)$, since multiplication is just the multiplication in the group G, and so elements of G are invertible in $\mathbf{Z}(G)$. Thus f restricted to G has image in $\mathrm{Un}(R)$ and this restriction is $\eta(f) \in \mathfrak{G}(G, \mathrm{Un}(R))$.

Now if $f \in \mathfrak{G}(G, \mathrm{Un}(R))$ it can be extended to a ring homomorphism $\mathbf{Z}(G) \to R$ in exactly one way, since G freely generates the additive structure of $\mathbf{Z}(G)$, and since f is a multiplicative homomorphism to start with. Thus η is an equivalence, and it is plainly natural. (The reader may prefer to postpone the contemplation of this example until he has read Section 1 of Chapter 4.)

5. Denote by $\mathfrak{S}_r$ the category whose objects are sets and whose morphisms are *relations*. If $R \subseteq X \times Y$ and $S \subseteq Y \times Z$, so that $R \in \mathfrak{S}_r(X, Y)$, $S \in \mathfrak{S}_r(Y, Z)$, we compose relations by the rule $SR \in \mathfrak{S}_r(X, Z)$, where

$SR = \{(x,z):$ there is $y \in Y$ such that $(x,y) \in R$ and $(y,z) \in S\}$. Let $P: \mathfrak{S}_r \to \mathfrak{S}$ be the covariant power set functor. That is, $PX = 2^X$, and if $R \subseteq X \times Y$, $PR: 2^X \to 2^Y$ is the function $PR(A) = \{y \in Y : (x,y) \in R$ for some $x \in A\}$, $A \subseteq X$. The reader should check that P is a functor. Now if Emb$: \mathfrak{S} \to \mathfrak{S}_r$ is the embedding functor, we see that Emb$\dashv P$; in other words

$$\mathfrak{S}_r(X,Y) \cong \mathfrak{S}(X, 2^Y).$$

That is, any relation from X to Y can be regarded as a function from X to 2^Y, in a natural way.

6. We now describe an interesting and exceptional case—a functor admitting both a left and a right adjoint. Let $U: \mathfrak{D}_K \to \mathfrak{Ab}$ be the underlying functor from the category $\mathfrak{D}_K$ of vector spaces over the field K to $\mathfrak{Ab}$. Denote by $F^r: \mathfrak{Ab} \to \mathfrak{D}_K$ the functor that takes A to the set $\mathrm{Hom}(K,A)$ of group homomorphisms from K to A, which we make into a vector space as follows. Since A is an abelian group, so is $\mathrm{Hom}(K,A)$. To give it a vector space structure, suppose that $f \in \mathrm{Hom}(K,A)$, $r \in K$, and define $fr: K \to A$ to be the function $(fr)(s) = f(rs)$, $s \in K$.

It is easy to check that fr is again a homomorphism and that $\mathrm{Hom}(K,A)$ becomes, by this rule, a vector space. Then $U \xrightarrow{\eta} F^r$. That is, if $M \in \mathfrak{D}_K$ and $A \in \mathfrak{Ab}$

$$\mathfrak{Ab}(UM, A) \stackrel{\eta}{\cong} \mathfrak{D}_K(M, \mathrm{Hom}(K,A)).$$

To see this, let $\phi: UM \to A$ be an abelian group homomorphism. Define $\phi': M \to \mathrm{Hom}(K,A)$ by the rule $\phi'(m)(r) = \phi(mr)$, $m \in M, r \in K$. Then ϕ' can be shown to be a K-homomorphism. Now let $\psi: M \to \mathrm{Hom}(K,A)$ be a K-homomorphism. Define $\psi': UM \to A$ by $\psi'(m) = \psi(m)(1)$. It is not difficult to show that the rules $\phi \mapsto \phi'$, $\psi \mapsto \psi'$ are mutual inverses and that the former is thus an adjugant $\eta: U \dashv F^r$.

Now define $F^l: \mathfrak{Ab} \to \mathfrak{D}_K$ by $F^l(A) = A \otimes K$. This is an abelian group which becomes a vector space by defining $(a \otimes r)s = a \otimes (rs)$ for $s \in K$. We now have $F^l \xrightarrow{\eta} U$; that is

$$\mathfrak{D}_K(A \otimes K, M) \stackrel{\eta}{\cong} \mathfrak{Ab}(A, UM).$$

To see this, let $\phi: A \otimes K \to M$ be a K-homomorphism. Define $\phi': A \to UM$ by $\phi'(a) = \phi(a \otimes 1)$. Conversely, if $\psi: A \to UM$ is an abelian group homomorphism, define $\psi': A \otimes K \to M$ by $\psi'(a \otimes r) = \psi(a)r, r \in K$. Then ψ' is a $K-$ homomorphism, and again we have mutually inverse natural transformations yielding the adjunction $F^l \dashv U$. Thus $F^l \dashv U \dashv F^r$.

We now proceed with the theory of adjoint functors.

THEOREM 7.1. *If* $F_1 \dashv G_1$ *and* $F_2 \dashv G_2$ *then* $F_1F_2 \dashv G_2G_1$.
The proof is left to the reader.

THEOREM 7.2. *A functor determines its right adjoint up to canonical natural equivalence.*

PROOF. We first explain the assertion. Suppose that $F: \mathfrak{C} \to \mathfrak{D}$, $G, G': \mathfrak{D} \to \mathfrak{C}$ and $F \overset{\eta}{\dashv} G$, $F \overset{\eta'}{\dashv} G'$. We must find a natural equivalence $\tau: G \to G'$ such that, for all $A \in \mathfrak{C}$ and $B \in \mathfrak{D}$, the following diagram commutes, where we have written τ for $\mathfrak{C}(A, \tau_B)$.

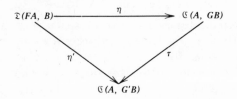

Fix $B \in \mathfrak{D}$. Then $\eta'\eta^{-1}: \mathfrak{C}(-, GB) \cong \mathfrak{C}(-, G'B)$. Here $\mathfrak{C}(-, GB), \mathfrak{C}(-, G'B)$ are contravariant functors $\mathfrak{C} \to \mathfrak{S}$. By Exercise 3.4, we know that this equivalence is achieved by a unique unit $\tau_B: GB \to G'B$. Now using the naturality of $\eta'\eta^{-1}$ in B we infer that τ is natural and $\eta'\eta^{-1} = \tau$.

We now insert a theorem that shows the importance of the concept of adjoint functors.

THEOREM 7.3. *If* G *has a left adjoint,* G *preserves products, kernels, pullbacks, and monics.*

PROOF. We are content to prove the assertion about pullbacks; the remaining assertions are proved similarly. Suppose that $F: \mathfrak{C} \to \mathfrak{D}$, $G: \mathfrak{D} \to \mathfrak{C}$, and $F \overset{\eta}{\dashv} G$. Suppose that

$$
\begin{array}{ccc}
P & \overset{\psi_1}{\to} & X_1 \\
{\scriptstyle \psi_2}\downarrow & & \downarrow{\scriptstyle \phi_1} \\
X_2 & \underset{\phi_2}{\to} & A
\end{array}
$$

is a pullback in $\mathfrak{D}$. We must show that

$$GP \xrightarrow{G\psi_1} GX$$
$$G\psi_2 \downarrow \qquad \downarrow G\phi_1$$
$$GX_2 \xrightarrow[G\phi_2]{} GA$$

is a pullback in $\mathfrak{C}$. Suppose that we have the following diagram in $\mathfrak{C}$.

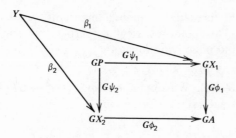

Applying η^{-1} and exploiting (7.3), we obtain the following commutative diagram in $\mathfrak{D}$:

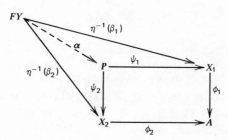

The existence of α then follows from the fact that we have a pullback in $\mathfrak{D}$. We now return to our original diagram, using η, and infer that the following diagram commutes:

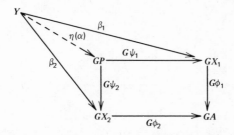

Moreover, $\eta(\alpha)$ is unique, because, if γ, say, also fitted into the diagram, then α and $\eta^{-1}(\gamma)$ would both fit into the diagram in $\mathfrak{D}$, so that $\alpha = \eta^{-1}(\gamma)$, $\eta(\alpha) = \gamma$. Thus we have a pullback in $\mathfrak{C}$ as claimed.

Note. In order to show that G preserves kernels, we should first show that G preserves zero. This is easy if we have first shown that G preserves products, for we have shown in Exercise 5.2 that, in any category with zero object, X is a zero object if and only if $(X; 1, 1)$ is the product of X and X. But then $(GX; 1, 1)$ is the product of GX and GX, and so GX is a zero object.

We now prepare for a different, but important, point of view about adjoint functors. This point of view is particularly suggested by the examples of free and underlying functors.

If $F \dashv G$ we know $\mathfrak{D}(FA, B) \overset{\eta}{\cong} \mathfrak{C}(A, GB)$. Let $B = FA$. Then $\mathfrak{D}(FA, FA) \overset{\eta}{\cong} \mathfrak{C}(A, GFA)$. We obtain $\epsilon_A = \eta(1_{FA}): A \to GF(A)$. It is plain that $\epsilon: 1 \to GF$ is a natural transformation,

$$\begin{array}{ccc} & \epsilon_{A_1} & \\ A_1 & \longrightarrow & GFA_1 \\ f\downarrow & & \downarrow GFf \\ A & \underset{\epsilon_A}{\longrightarrow} & GFA \end{array}$$

For $GFf \circ \epsilon_{A_1} = GFf \circ \eta(1_{FA_1}) = \eta(Ff \circ 1_{FA_1}) = \eta(Ff) = \eta(1_{FA} \circ Ff) = \eta(1_{FA}) \circ f = \epsilon_A \circ f$. We call ϵ the *unit* or *front* of the adjunction.

Now let $A = GB$. Then $\mathfrak{D}(FGB, B) \overset{\eta}{\cong} \mathfrak{C}(GB, GB)$. Let $\delta_B = \eta^{-1}(1_{GB}): FGB \to B$. Then $\delta: FG \to 1$ is a natural transformation called the *counit* or *rear* of the adjunction.

THEOREM 7.4.

(i)

$$\begin{array}{ccc} & F\epsilon & \delta F \\ F & \longrightarrow FGF & \longrightarrow F \\ & \underset{1}{\underline{\qquad\qquad}} \nearrow & \end{array} \tag{7.4}$$

(ii)

$$\begin{array}{ccc} & \epsilon G & G\delta \\ G & \longrightarrow GFG & \longrightarrow G \\ & \underset{1}{\underline{\qquad\qquad}} \nearrow & \end{array}$$

PROOF. To prove Theorem 7.4(i), we must show that $\delta_{FA} \circ F(\epsilon_A) = 1: FA \to FA$; but this is an immediate consequence of (7.3) and the definitions of ϵ, δ. Similarly, we prove Theorem 7.4(ii).

Notice that if we are given ϵ_A or δ_B as defined above, we can recover η. Suppose that $f: FA \rightarrow B$, and $g: A \rightarrow GB$. Then it is easy to prove, again using (7.3), that $\eta(f) = Gfo\epsilon_A$, and $\eta^{-1}(g) = \delta_B oFg$. We now enunciate an important converse to what has been stated above.

THEOREM 7.5. *Given functors* $F: \mathfrak{C} \rightarrow \mathfrak{D}, G: \mathfrak{D} \rightarrow \mathfrak{C}$ *and natural transformations* $\epsilon: 1 \rightarrow GF$ *and* $\delta: FG \rightarrow 1$ *such that the identities* (7.4)(i) *and* (ii) *hold, the rule* $\eta(f) = Gfo\epsilon_A$, *for* $f: FA \rightarrow B$, *defines an adjunction* $F \overset{\eta}{\dashv} G$ *such that* ϵ *and* δ *are the front and rear of the adjunction.*

PROOF. Certainly η is natural. Now define $\bar{\eta}: \mathfrak{C}(A, GB) \rightarrow \mathfrak{D}(FA, B)$ by $\bar{\eta}(g) = \delta_B oFg$. Then $\bar{\eta}\eta(f) = \delta_B oFGfoF\epsilon_A = fo\delta_{FA} oF\epsilon_A$, by the naturality of δ, so that $\bar{\eta}\eta(f) = f$, by (7.4)(i). Similarly, $\eta\bar{\eta} = 1$. The front of the adjunction ϵ' is then given by $\epsilon'_A = \eta(1_{FA}) = \epsilon_A$. Thus $\epsilon' = \epsilon$; and similarly the rear of the adjunction is δ.

As an example of the front of an adjunction consider $Fr: \mathfrak{S} \rightarrow \mathfrak{Ab}$, the free functor. Then $Fr \dashv U$. If $S \in \mathfrak{S}$ then $\epsilon: S \rightarrow U(Fr(S))$ is the embedding of S in $U(Fr(S))$, that is, the embedding of the set S as the set of free generators, or basis, of the free abelian group $Fr(S)$.

As an example of the rear of an adjunction, consider $U: \mathfrak{Ab} \rightarrow \mathfrak{S}$, the underlying functor. If $A \in \mathfrak{Ab}$, $\delta_A: Fr(U(A)) \rightarrow A$ is the canonical homomorphism which is the identity on the basis A of $Fr(U(A))$. This gives the canonical free abelian presentation of A. Similarly, for $U: \mathfrak{G} \rightarrow \mathfrak{S}$, δ_G is the canonical free presentation of the group G.

Consider now the third example we gave of a pair of adjoint functors:

$$\text{Abel} \overset{\eta}{\dashv} U,$$

where $\text{Abel}: \mathfrak{G} \rightarrow \mathfrak{Ab}$ is the abelianizing functor, and $U: \mathfrak{Ab} \rightarrow \mathfrak{G}$ is the underlying functor, which may also be thought of as an embedding functor. If G is a group and A an abelian group, then any homomorphism $\phi: G \rightarrow A$ determines a homomorphism $\bar{\phi}: G/G' \rightarrow A$, since $G' \subseteq \ker\phi$, and

$$\eta(\phi) = \bar{\phi}.$$

In this case, the rear of the adjunction is just the identity, $\delta_A = 1_A: A \rightarrow A$. This phenomenon (where the front or rear of an adjunction is the identity) is of particular importance and we now examine the circumstances that give rise to it.

THEOREM 7.6. *Let* $F: \mathfrak{C} \rightarrow \mathfrak{D}, G: \mathfrak{D} \rightarrow \mathfrak{C}, F \dashv G$, *and let* G *be full and faithful. Then the rear* δ *of the adjunction is a natural equivalence.*

PROOF. Consider $\epsilon_{GB}: GB \rightarrow GFGB$. Since G is full and faithful, there exists a unique $\rho_B : B \to FGB$ such that $G\rho_B = \epsilon_{GB}$, and it is easy to see that $\rho : 1 \to FG$ is a natural transformation. We will show that ρ is inverse to δ. First, by (7.4)(ii)

$$G\delta_B \, o \, G\rho_B = G\delta_B \, o \, \epsilon_{GB} = 1,$$

so that $\delta\rho = 1, G$ being faithful. Second,

$$\eta(\rho_B \, o \, \delta_B) = G\rho_B \, o \, \eta(\delta_B) = G\rho_B = \epsilon_{GB} = \eta(1_{FGB}),$$

so that $\rho\delta = 1$.

THEOREM 7.7. *Let* $F : \mathfrak{C} \to \mathfrak{D}, G : \mathfrak{D} \to \mathfrak{C}, F \dashv G$, *and let* G *be a full embedding. Then we may choose a new left adjoint* F' *to* G *such that the rear of the adjunction* $\eta' : F' \dashv G$ *is the identity.*

PROOF. We use $\rho : 1 \to FG$ as in the proof of Theorem 7.6 and define $F' : \mathfrak{C} \to \mathfrak{D}$ as follows:

$$F'A = FA \qquad \text{if } A \notin \text{im } G,$$

$$F'GB = B,$$

and, if $\alpha : A_1 \to A_2$ in $\mathfrak{C}$,

$$F'\alpha = F\alpha \qquad \text{if } A_1, A_2 \notin \text{im } G,$$

$$= \beta \qquad \text{if } A_1, A_2 \in \text{im } G \text{ and } \alpha = G\beta,$$

$$= F\alpha \, o \, \rho_{B_1} \qquad \text{if } A_1 = GB_1, F_2 \notin \text{im } G,$$

$$= \delta_{B_2} \, o \, F\alpha \qquad \text{if } A_1 \notin \text{im } G, A_2 = GB_2.$$

Notice that these definitions make sense, since G is injective on objects.

We leave it to the reader to check that F' is a functor. We have a natural equivalence of functors $\tau : F \cong F'$, given by $\tau_A = 1_{FA}$ if $A \notin \text{im } G$, $\tau_A = \delta_B$ if $A = GB$. Thus $F' \xrightarrow{\eta'} G$, where $\eta'(f') = \eta(f' \, o \, \tau_A), f' : F'A \to B$; and the rear δ' of this adjunction is the identity, since

$$\eta'(1_B) = \eta(\tau_{GB}) = \eta(\delta_B) = 1_{GB} = \eta'(\delta_B').$$

8. ABELIAN CATEGORIES

In this section we consider a class of categories that includes $\mathfrak{Ab}$ and the categories of modules studied in Chapter 4.

DEFINITION 8.1. An *additive category* $\mathfrak{A}$ is a category with zero object and finite products, such that, for every $A, B \in A$, the set $\mathfrak{A}(A, B)$ has a canonical abelian group structure. That is, each $\mathfrak{A}(A, B)$ is an abelian group, and

$$(f + g)h = fh + gh, f, g \in \mathfrak{A}(B, C), h \in \mathfrak{A}(A, B),$$

$$f(g + h) = fg + fh, f \in \mathfrak{A}(B, C), g, h \in \mathfrak{A}(A, B). \tag{8.1}$$

Examples are $\mathfrak{A}\mathfrak{b}$ and $\mathfrak{D}_K$. We remark that it follows easily from (8.1) that the zero morphism 0_{AB} is the zero of the group $\mathfrak{A}(A, B)$. Note also that (8.1) may be interpreted as saying that the functors $\mathfrak{A}(A, -), \mathfrak{A}(-, B)$ (see Exercises 2.7 and 4.7) are functors to $\mathfrak{A}\mathfrak{b}$ instead of $\mathfrak{S}$.

We will show that the axioms of an additive category are really self-dual. This clearly amounts to showing that $\mathfrak{A}$ has finite coproducts. In fact, we prove a more precise result. Suppose that $(A_1 \oplus A_2; p_1, p_2)$ is the product of A_1 and A_2. Then define $q_1 = \{1, 0\} : A_1 \to A_1 \oplus A_2$ and $q_2 = \{0, 1\} : A_2 \to A_1 \oplus A_2$.

THEOREM 8.1. $(A_1 \oplus A_2; q_1, q_2)$ *is the coproduct of* A_1 *and* A_2 *in* A.

PROOF. We show first that $q_1 p_1 + q_2 p_2 : A_1 \oplus A_2 \to A_1 \oplus A_2$ is the identity. Now $p_i q_j = 1$ if $i = j$ and $p_i q_j = 0$ if $i \neq j$. Thus $p_1(q_1 p_1 + q_2 p_2) = p_1 q_1 p_1 + p_1 q_2 p_2 = p_1 + 0 = p_1$. Similarly $p_2(q_1 p_1 + q_2 p_2) = p_2$. Thus $q_1 p_1 + q_2 p_2 = \{p_1, p_2\} = 1$. Returning to the theorem, we must find α such that

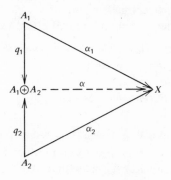

and prove it unique. Let $\alpha = \alpha_1 p_1 + \alpha_2 p_2$. Then $\alpha q_i = (\alpha_1 p_1 + \alpha_2 p_2) q_i = \alpha_1 p_1 q_i + \alpha_2 p_2 q_i = \alpha_i, i = 1, 2$. To show that α is unique, suppose that $\alpha q_i = \alpha' q_i, i = 1, 2$. Then, by the first step in the proof of the theorem,

$$\alpha = \alpha(q_1 p_1 + q_2 p_2) = \alpha q_1 p_1 + \alpha q_2 p_2 = \alpha' q_1 p_1 + \alpha' q_2 p_2 = \alpha'(q_1 p_1 + q_2 p_2) = \alpha'.$$

In an additive category, we usually refer to the coproduct as the *sum*. Notice that the sum object coincides with the product object, for any finite collection of objects of $\mathfrak{A}$. We next show

THEOREM 8.2.

$$A \xrightarrow[\ \ \ \ g_1f_1 + g_2f_2\ \ \ \ ]{\{f_1,f_2\}} B_1 \oplus B_2 \xrightarrow{\langle g_1,g_2 \rangle} C$$

PROOF. Notice that, in the proof of Theorem 8.1, $\alpha = \langle \alpha_1, \alpha_2 \rangle = \alpha_1 p_1 + \alpha_2 p_2$. Thus $\langle g_1, g_2 \rangle \{ f_1, f_2 \} = (g_1 p_1 + g_2 p_2) \{ f_1, f_2 \} = g_1 p_1 \{ f_1, f_2 \} + g_2 p_2 \{ f_1, f_2 \} = g_1 f_1 + g_2 f_2$.

COROLLARY 8.3. *The group operation in each $\mathfrak{A}(A,B)$ is uniquely determined by the additive category $\mathfrak{A}$.*

PROOF. Set $g_1 = g_2 = 1$ above. Then $f_1 + f_2 = \langle 1, 1 \rangle \{ f_1, f_2 \}$, so that $f_1 + f_2$ is determined by the product (coproduct) in $\mathfrak{A}$.

Naturally, since we are here concerned with categories with additional structure, we give special attention to structure-preserving functors. We make the following definition, but immediately prove that our definition yields functors that genuinely are structure-preserving.

DEFINITION 8.2. An *additive functor* $F : \mathfrak{A} \to \mathfrak{B}$ between additive categories is a product-preserving functor.

To justify this rather odd-sounding definition, we now prove

PROPOSITION 8.4. *The following statements about a functor $F : \mathfrak{A} \to \mathfrak{B}$ of additive categories are equivalent:*

(*i*) F *preserves finite products.*
(*ii*) F *preserves finite coproducts (sums).*
(*iii*) $F(f + g) = Ff + Fg$.

PROOF. That (i) holds if and only if (ii) holds follows from Theorem 8.1, once one recalls that if F preserves products it must preserve zero morphisms (see Exercise 5.2). To show that (i) and (ii) imply (iii) consider

$$A \xrightarrow[\ \ \ \ f + g\ \ \ \ ]{\{1,1\}} A \oplus B \xrightarrow{\langle f,g \rangle} B$$

Then $F(f + g) = F(\langle f, g \rangle \{1, 1\}) = F\langle f, g \rangle F\{1, 1\} = \langle Ff, Fg \rangle \{1, 1\} = Ff + Fg$. To show that (iii) implies (i) consider

$$F(A_1 \oplus A_2) \xrightarrow{\{Fp_1, Fp_2\}} FA_1 \oplus FA_2.$$

We must show that $F(A_1 \oplus A_2)$ is a product, and this amounts to showing that $\{Fp_1, Fp_2\}$ is a unit. We show that its inverse is

$$FA_1 \oplus FA_2 \xrightarrow{(Fq_1)p_1 + (Fq_2)p_2} F(A_1 \oplus A_2).$$

Now

$$((Fq_1)p_1 + (Fq_2)p_2)\{Fp_1, Fp_2\} = (Fq_1)p_1\{Fp_1, Fp_2\} + (Fq_2)p_2\{Fp_1, Fp_2\}$$

$$= (Fq_1)(Fp_1) + (Fq_2)(Fp_2)$$

$$= F(q_1p_1 + q_2p_2), \text{ by (iii)},$$

$$= F1$$

$$= 1.$$

On the other hand,

$$\{Fp_1, Fp_2\}((Fq_1)p_1 + (Fq_2)p_2) = \{Fp_1, Fp_2\}(Fq_1)p_1 + \{Fp_1, Fp_2\}(Fq_2)p_2$$

$$= \{(Fp_1)(Fq_1)p_1, (Fp_2)(Fq_1)p_1\} + \{(Fp_1)(Fq_2)p_2, (Fp_2)(Fq_2)p_2\}$$

$$= \{p_1, 0\} + \{0, p_2\}, \text{ since } F(0) = 0$$

$$= q_1p_1 + q_2p_2$$

$$= 1.$$

Thus any of the criteria (i), (ii), and (iii) above can be used to test if F is additive.

Before defining the notion of an abelian category, we make one more useful remark about additive categories. To test whether a morphism in an additive category is monic (epic) we simply have to show that its kernel (cokernel) is zero. Notice that this may not be the case in a given category with zero. For example, in $\mathfrak{G}$ the statement about epics is, in fact, false.

DEFINITION 8.3. An *abelian category* is an additive category in which
   (i) every morphism has a kernel and a cokernel,
   (ii) every monic (epic) is the kernel (cokernel) of its cokernel (kernel),
   (iii) every morphism is expressible as an epic followed by a monic,
The axioms are self-evidently self-dual!

Examples of abelian categories are $\mathfrak{Ab}, \mathfrak{D}_K$, graded abelian groups, torsion abelian groups, and finite abelian groups.

An example of a category that is additive but not abelian is the category $\mathfrak{F}$ of free abelian groups; but beware of the following argument. Denote by $\mathbf{Z} \xrightarrow{2} \mathbf{Z}$ the morphism $n \mapsto 2n$. In $\mathfrak{Ab}$ we have

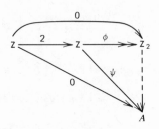

where $\phi: \mathbf{Z} \twoheadrightarrow \mathbf{Z}_2$ is the projection, and so the cokernel of $\mathbf{Z} \xrightarrow{2} \mathbf{Z}$ in $\mathfrak{Ab}$ is $\mathbf{Z} \xrightarrow{\phi} \mathbf{Z}_2$. Since $\mathbf{Z}_2$ is not free, this cannot be the cokernel in $\mathfrak{F}$. We must be careful, however, not to conclude that there is no cokernel in $\mathfrak{F}$, for there is. It is $\mathbf{Z} \to 0$.

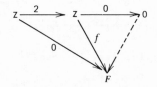

Here, $F \in \mathfrak{F}$, so that if $2f = 0, f = 0$. Thus the cokernel of $\mathbf{Z} \xrightarrow{2} \mathbf{Z}$ in $\mathfrak{F}$ is $\mathbf{Z} \xrightarrow{0} 0$, that is, $\mathbf{Z} \xrightarrow{2} \mathbf{Z}$ is epic in $\mathfrak{F}$. But the kernel of $\mathbf{Z} \xrightarrow{0} 0$ is $\mathbf{Z} \xrightarrow{1} \mathbf{Z}$, which is not equivalent to $\mathbf{Z} \xrightarrow{2} \mathbf{Z}$, since the latter is not a unit of $\mathfrak{F}$. On the other hand, $\mathbf{Z} \xrightarrow{2} \mathbf{Z}$ is certainly monic in $\mathfrak{F}$. Thus (ii) is contradicted. (It is an interesting question whether all morphisms in $\mathfrak{F}$ have cokernels.)

In any category with zero, finding the kernel of $\xrightarrow{\eta}$ is the same as finding the kernel of $\xrightarrow{\eta} \rightarrowtail$. In abelian categories we have a partial converse. We suppose henceforth in this section that our categories are abelian.

LEMMA 8.5. *If η is epic and η and $\gamma\eta$ have the same kernel, then γ is monic.*

PROOF. Use (iii) to set $\gamma = \mu\epsilon$ where ϵ is epic and η is monic. Then $\gamma\eta = \mu\epsilon\eta$ and so, by the remark above,

$$\ker \eta = \ker \gamma\eta = \ker \epsilon\eta.$$

By property (ii), η and $\epsilon\eta$ are both cokernels of a common monic (ker η). Thus there is a unit u such that $u\eta = \epsilon\eta$. Since η is epic, $u = \epsilon$ and ϵ therefore is a unit. Thus $\gamma = \mu\epsilon$ is monic.

Consider a morphism α in an abelian category $\mathfrak{A}$. We have $\overset{\mu}{\rightarrowtail} \overset{\alpha}{\rightarrow} \overset{\epsilon}{\twoheadrightarrow}$ where $\mu = \ker\alpha$ and $\epsilon = \operatorname{coker}\alpha$. If $\eta = \operatorname{coker}\mu$ we have a unique γ such that

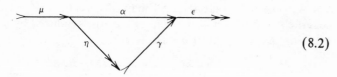

$$(8.2)$$

Thus μ is the kernel of $\gamma\eta$ and of η so that γ is monic, by Lemma 8.5. Now ϵ is the cokernel of α and so ϵ is the cokernel of $\gamma\eta$. Thus, in (8.2), since η is epic, ϵ is the cokernel of γ and γ is the kernel of ϵ. To express all these relations symmetrically let us introduce some notation. Thus

$$f]g \text{ means } f \text{ is the kernel of } g,$$

$$f[\, g \text{ means } g \text{ is the cokernel of } f,$$

$$f\square g \text{ means } f]g \text{ and } f[\, g.$$

Summing up, then, we have proved that $\mu]\alpha$ and $\alpha[\epsilon$ imply that $\alpha = \gamma\eta$, with $\mu\square\eta$ and $\gamma\square\epsilon$. This shows that the factorization in (iii) of a morphism into an epic followed by a monic is essentially unique. For if $\alpha = \gamma\eta, \eta$ epic, γ monic, then $\eta = \operatorname{coker}(\ker\alpha)$, $\gamma = \ker(\operatorname{coker}\alpha)$. Thus

THEOREM 8.6. *If* $\alpha = \gamma\eta = \gamma'\eta'$, *with* γ, γ' *monic,* η, η' *epic, then there is a* (*unique*) *unit* u *such that*

$$\eta' = u\eta, \gamma'u = \gamma.$$

The *factorization* $\overset{\mu}{\rightarrowtail} \cdot \overset{\eta}{\twoheadrightarrow} \cdot \overset{\gamma}{\rightarrowtail} \cdot \twoheadrightarrow \epsilon$, where $\alpha = \gamma\eta$ is plainly functorial in the sense that, given a commutative diagram

$$
\begin{array}{c}
\overset{\alpha}{\rightarrow} \\
\phi\downarrow \qquad \downarrow\psi \\
\underset{\alpha'}{\rightarrow}
\end{array}
$$

we obtain a unique map of factorizations

$$\begin{array}{ccccccc}
\cdot \overset{\mu}{\rightarrowtail} & \cdot & \overset{\eta}{\twoheadrightarrow} & \cdot & \overset{\gamma}{\rightarrowtail} & \cdot & \overset{\epsilon}{\twoheadrightarrow} & \cdot \\
\downarrow \rho & \downarrow \phi & & \downarrow \sigma & \downarrow \psi & & \downarrow \tau & (\alpha=\gamma\eta, \alpha'=\gamma'\eta') \\
\cdot \underset{\mu'}{\twoheadrightarrow} & \cdot & \underset{\eta'}{\twoheadrightarrow} & \cdot & \underset{\gamma'}{\rightarrowtail} & \cdot & \underset{\epsilon'}{\twoheadrightarrow} & \cdot
\end{array}$$

For we obtain ρ as the induced map of kernels, τ as the induced map of cokernels, and then σ as the map of cokernels induced by ρ. It is then only necessary to verify that $\gamma'\sigma = \psi\gamma$. But

$$\gamma'\sigma\eta = \gamma'\eta'\phi = \alpha'\phi = \psi\alpha = \psi\gamma\eta,$$

and η is epic.

THEOREM 8.7. *In an abelian category, a morphism is a unit if and only if it is monic and epic.*

PROOF.   Suppose that α is monic and epic. Then $\alpha = \alpha 1$, $\alpha = 1\alpha$ are both factorizations of α into an epic followed by a monic. But then there is a unit u such that $\alpha = 1u = u$, that is, α is a unit.

Of course the conclusion of Theorem 8.7 is valid in many nonabelian categories (for example, $\mathfrak{S}$ and $\mathfrak{G}$). However, it is extremely important to note that it is in general false (for example, in $\mathfrak{T}$). One certainly should never *define* an isomorphism to be a morphism which is epic and monic. (This is, unfortunately, the habit in many textbooks on group theory.)

We now proceed to define one of the most important concepts of abelian category theory.

DEFINITION 8.4.   The sequence $\cdots \overset{f}{\rightarrow} A \overset{g}{\rightarrow} \cdots$ is *exact at A* if, when $f = \mu_f \epsilon_f$ and $g = \mu_g \epsilon_g$ are the factorizations of Definition 8.3(iii), then $\mu_f \square \epsilon_g$. The sequence $\cdots \rightarrow \cdot \rightarrow \cdots$ is *exact* if it is exact at every point.

In other words, μ_f is the kernel of g; or equivalently, ϵ_g is the cokernel of f. However, our definition has the advantage, over either of these formulations, of being symmetrical.

Notice that in $\mathfrak{Ab}$ this notion of exactness means that the image of f coincides with the kernel of g, in the traditional sense. Thus Definition 8.4 provides a generalization of the notion of exactness presented in Chapter 2.

Some exact sequences have special names.

$$\cdot \rightarrow \cdot \twoheadrightarrow \cdot \qquad \textit{right exact,}$$

$$\cdot \rightarrowtail \cdot \rightarrow \cdot \qquad \textit{left exact,} \tag{8.3}$$

$$\cdot \rightarrowtail \cdot \twoheadrightarrow \cdot \qquad \textit{short exact.}$$

Consider now the exact sequence

$$A' \xrightarrow{\alpha'} A \xrightarrow{\alpha''} A'' \to 0. \tag{8.4}$$

Exactness at A'' means that $A'' \to 0$ is the cokernel of $A \xrightarrow{\alpha''} A''$. But since α'' has a zero cokernel, α'' is epic; thus (8.4) is a right exact sequence $A' \xrightarrow{\alpha'} A \xrightarrow{\alpha''} A''$. Exactness at A means that $A \xrightarrow{\alpha''} A''$ is the cokernel of $A' \xrightarrow{\alpha'} A$. We therefore have the following interpretations (using the duality principle).

THEOREM 8.8. *The sequence* $A' \xrightarrow{\alpha'} A \xrightarrow{\alpha''} A''$ *is*
  (i) *right exact if and only if* $\alpha'[\alpha''$,
  (ii) *left exact if and only if* $\alpha']\alpha''$,
  (iii) *short exact if and only if* $\alpha' \,\Box\, \alpha''$.

DEFINITION 8.5. A functor F from an abelian category $\mathfrak{A}$ to an abelian category $\mathfrak{B}$ is *left exact* if it takes left exact sequences to left exact sequences. We similarly define a *right exact* functor. A functor is *exact* if it takes short exact sequences to short exact sequences.

Suppose that we have $A' \xrightarrow{\alpha'} A \xrightarrow{\alpha''} A''$ left exact. Applying the functor $\mathfrak{A}(X,-):\mathfrak{A} \to \mathfrak{A}\mathfrak{b}$ we get $\mathfrak{A}(X,A') \xrightarrow{\alpha'_*} \mathfrak{A}(X,A) \xrightarrow{\alpha''_*} \mathfrak{A}(X,A'')$. Now α' monic means that α'_* is injective; because, recalling the definition of $\alpha'_* = \mathfrak{A}(X,\alpha')$ as $\alpha'_*(f) = \alpha'f$, we have $\alpha'_*(f) = \alpha'_*(g)$ if and only if $\alpha'f = \alpha'g$, so that α' is monic if and only if α'_* is injective. Thus we have

$$\mathfrak{A}(X,A') \rightarrowtail \mathfrak{A}(X,A) \xrightarrow{\alpha''_*} \mathfrak{A}(X,A'').$$

Similarly, $\alpha']\alpha''$ means that α'_* is the kernel of α''_*. To see this, notice that $f \in \mathfrak{A}(X,A)$ is in the kernel of α''_* if and only if $\alpha''f = 0$. But then if $\alpha']\alpha''$, there is g such that $\alpha'g = f$. This means that $\alpha'_*(g) = f$, and f is in the image of α'_*. Thus α'_* is the kernel of α''_*. Conversely, if $\alpha'_* = \ker\alpha''_*$, then $\alpha''f = 0$ means that $f = \alpha'_*(g)$ for some $g:X \to A'$; that is, $f = \alpha'g$, and $\alpha']\alpha''$. Thus we have proved (i) below; (ii) follows similarly.

THEOREM 8.8.
  (i) $A' \to A \to A''$ *is left exact if and only if* $Hom(X,A') \to Hom(X,A) \to Hom(X,A'')$ *is left exact for all* X.
  (ii) $A' \to A \to A''$ *is right exact if and only if* $Hom(A'',X) \to Hom(A,X) \to Hom(A',X)$ *is left exact for all* X.
  Here we use the notation $\mathrm{Hom}(A,B)$ for $\mathfrak{A}(A,B)$. This is standard in an abelian category.

THEOREM 8.9. *A functor (of abelian categories) with a right adjoint preserves right exact sequences, that is, it is right exact.*

PROOF. Suppose that $F:\mathfrak{C}\to\mathfrak{D}$, $G:\mathfrak{D}\to\mathfrak{C}$ and $F\dashv G$. Suppose that $A'\to A\to A''$ is right exact. We must prove that $FA'\to FA\to FA''$ is right exact. By Theorem 8.8(ii) this is equivalent to saying the following is left exact, for all X:

$$\mathrm{Hom}\,(FA'',X)\to\mathrm{Hom}\,(FA,X)\to\mathrm{Hom}\,(FA',X).$$

Since $F\dashv G$, this is equivalent to the following being left exact for all X:

$$\mathrm{Hom}\,(A'',GX)\to\mathrm{Hom}\,(A,GX)\to\mathrm{Hom}\,(A',GX).$$

But $A'\to A\to A''$ right exact implies that the above is left exact, again by Theorem 8.8(ii).

As an application, it will be shown (Theorem 4.3 of Chapter 4) that the tensor product $-\otimes B:\mathfrak{Ab}\to\mathfrak{Ab}$ has the Hom functor $\mathrm{Hom}(B,-):\mathfrak{Ab}\to\mathfrak{Ab}$ as its right adjoint. *Thus the tensor product is right exact.* The reader should compare this proof, based on what the tensor product *does*, with the classical proof, based on what the tensor product *is* (see Theorem 5.8 of Chapter 2).

THEOREM 8.10. *A functor (of abelian categories) is additive if it is right, or left, exact.*

PROOF. It will be sufficient to suppose $F:\mathfrak{C}\to\mathfrak{D}$ right exact. Let $A=A'\oplus A''$ in $\mathfrak{D}$ so that we have a short exact sequence $A'\xrightarrow{\iota'} A\xrightarrow{\pi''} A''$, together with a morphism $\pi':A\to A'$ such that $\pi'\iota'=1$. Then $FA'\xrightarrow{F\iota'} FA\xrightarrow{F\pi''} FA''$ is right exact, and we have a morphism $F\pi':FA\to FA'$ such that $(F\pi')(F\iota')=1$. But it is easy to see that this forces $FA=FA'\oplus FA''$, so that F is additive (see Definition 8.7 and the subsequent discussion).

We have defined a functor as right exact if it preserves right exact sequences. Note that such a functor preserves epics as it preserves the right exact sequence $\cdot\to\cdot\to 0$. Since any short exact sequence $\rightarrowtail\cdot\twoheadrightarrow$ is right exact, it gets sent by a right exact functor to a right exact sequence. It is interesting to note that this property characterizes right exact functors. That is, if F takes short exact sequences to right exact sequences, it is right exact. This is because a right exact sequence $\xrightarrow{f}\xrightarrow{g}\twoheadrightarrow$ can be factored as

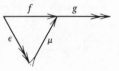

with $\mu \Box g$. Then F takes this to

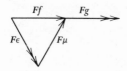

Let $\gamma \Box \epsilon$. Since F takes short exact sequences to right exact sequences, $F\gamma[F\epsilon$, so, in particular, $F\epsilon$ is epic. Thus, since $F\mu[Fg$, it follows that $(F\mu)(F\epsilon)[Fg$, so that $Ff[Fg$, and $\cdot \overset{Ff}{\to} \cdot \overset{Fg}{\twoheadrightarrow} \cdot$ is right exact.

In $\mathfrak{A}\,\mathfrak{b}$ consider the short exact sequence

$$\mathbf{Z} \overset{2}{\rightarrowtail} \mathbf{Z} \twoheadrightarrow \mathbf{Z}_2. \tag{8.5}$$

Now $\mathrm{Hom}(\mathbf{Z}_2, \mathbf{Z}) = \{0\}$ and $\mathrm{Hom}(\mathbf{Z}_2, \mathbf{Z}_2) = \mathbf{Z}_2$; thus, applying the left exact functor $\mathrm{Hom}(\mathbf{Z}_2, -)$, we get the left exact sequence

$$0 \rightarrowtail 0 \to \mathbf{Z}_2.$$

Thus $\mathrm{Hom}(\mathbf{Z}_2, -)$ is not an exact functor. Applying the right exact functor $\mathbf{Z}_2 \otimes -$ to the same short exact sequence (8.5), we get the right exact sequence

$$\mathbf{Z}_2 \overset{0}{\to} \mathbf{Z}_2 \rightarrowtail\!\!\!\twoheadrightarrow \mathbf{Z}_2.$$

Thus $\mathbf{Z}_2 \otimes -$ is not an exact functor. These examples lead to the following very natural question.

If we start with a short exact sequence $A' \rightarrowtail A \twoheadrightarrow A''$ and apply a left exact functor G we get a left exact sequence $GA' \rightarrowtail GA \to GA''$. We may ask: What is the cokernel of $GA \to GA''$? If we apply a right exact functor F we get a right exact sequence $FA' \to FA \twoheadrightarrow FA''$. We may ask: What is the kernel of $FA' \to FA$? These questions lead into the theory of *derived functors* [4]. Here we touch on two special cases (see also Chapter 7).

Given a short exact sequence $A' \rightarrowtail A \twoheadrightarrow A''$ we know $\mathrm{Hom}(X, A') \twoheadrightarrow \mathrm{Hom}(X, A) \to \mathrm{Hom}(X, A'')$ is left exact for every X. It is natural to ask for what X we can say it is actually short exact.

DEFINITION 8.6. $X \in \mathfrak{A}$ is *projective* if $\mathrm{Hom}(X, -)$ is an exact functor. Asking whether X is projective thus amounts to asking if there is always an h to complete the following diagram.

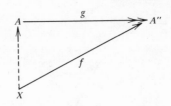

In $\mathfrak{Ab}$, we recall that, if X is free abelian, it is projective. For if $X = \mathrm{Fr}(S)$ we take $h(s)$ to be any counterimage of $f(s)$ under g, $s \in S$. The only projectives in $\mathfrak{Ab}$ are the free abelian groups. This follows from the fact that subgroups of free abelian groups are free abelian; for it can be shown that, in any abelian category, $X = X_1 \oplus X_2$ is projective if and only if X_1 and X_2 are projective.

We now introduce the "dual" notion.

DEFINITION 8.6*. $Y \in \mathfrak{A}$ is *injective* if $\mathrm{Hom}(-, Y)$ is an exact functor. Of course, $\mathrm{Hom}(-, Y)$ is always left exact. The appropriate diagram is, in this case,

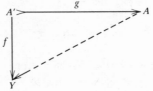

It is much harder to find injective objects of $\mathfrak{Ab}$ than projective objects. We call an abelian group A *divisible* if, given $a \in A$ and n a nonzero integer, there exists $b \in A$ such that $nb = a$. Then an abelian group Y is injective if and only if it is divisible. The reader is referred to Section 3 of Chapter 2.

DEFINITION 8.7. The short exact sequence $A' \overset{\alpha'}{\rightarrowtail} A \overset{\alpha''}{\twoheadrightarrow} A''$ *splits* if there exists $f: A'' \to A$ such that $f\alpha'' = 1$.

It can be shown (see Exercise 8.1) that an equivalent formulation of this definition is: There exists

$$g: A \to A' \quad \text{such that} \quad \alpha'g = 1.$$

Suppose now that we start with a fixed short exact sequence

$$A' \overset{\alpha'}{\rightarrowtail} A \overset{\alpha''}{\twoheadrightarrow} A''. \tag{8.6}$$

It is natural to ask what can be said if we know that

$$\mathrm{Hom}(X, A') \twoheadrightarrow \mathrm{Hom}(X, A) \to \mathrm{Hom}(X, A'') \tag{8.7}$$

is short exact for all X. In particular, we then know that $\mathrm{Hom}(X,A)$ $\twoheadrightarrow \mathrm{Hom}(X,A'')$, all X. Thus $\mathrm{Hom}(A'',A) \twoheadrightarrow \mathrm{Hom}(A'',A'')$, so that there is $f : A'' \to A$ such that $f\alpha'' = 1$. This simply means that the short exact sequence splits; indeed,

$$\langle \alpha',f \rangle : A' \oplus A'' \cong A.$$

It is obvious, conversely, that if (8.6) splits, then (8.7) is short exact—and splits.

Similarly, to say $\mathrm{Hom}(Y,A'') \twoheadrightarrow \mathrm{Hom}(Y,A) \to \mathrm{Hom}(Y,A')$ is short exact for all Y means that $\mathrm{Hom}(Y,A) \twoheadrightarrow \mathrm{Hom}(Y,A')$. In particular, $\mathrm{Hom}(A',A) \twoheadrightarrow \mathrm{Hom}(A',A')$ and so there is $g : A' \to A$ such that $\alpha'g = 1$. Again, this means that the short exact sequence (8.6) splits; indeed,

$$\{ g,\alpha'' \} : A \cong A' \oplus A''.$$

The converse is once more obvious. Thus we have proved

THEOREM 8.11. *Let*

$$\Sigma : A' \overset{\alpha'}{\rightarrowtail} A \overset{\alpha''}{\twoheadrightarrow} A''$$

be a short exact sequence in the abelian category $\mathfrak{Ab}$. *Then the following statements are equivalent*:

  (*i*) Σ *splits*,
  (*ii*) $\mathrm{Hom}(X,-)$ *transforms* Σ *into a short exact sequence for all* X *in* $\mathfrak{Ab}$;
  (*iii*) $\mathrm{Hom}(-,Y)$ *transforms* Σ *into a short exact sequence for all* Y *in* $\mathfrak{Ab}$.

We note that X is projective if and only if $\mathrm{Hom}(X,A) \to \mathrm{Hom}(X,A'')$ is epic for all $A \twoheadrightarrow A''$. This, of course, can be asked in any category, not only abelian categories. Apparently, then, we can define projective (and injective) objects in any category. We therefore allow ourselves a digression in this direction.

DEFINITION 8.8. An object $X \in \mathfrak{C}$ is *projective* if $\mathfrak{C}(X,-)$ preserves epics. An object X in *injective* if $\mathfrak{C}(-,X)$ transforms monics to epics.

In $\mathfrak{S}$, all objects are both projective and injective. We use this fact to show why free objects have always turned out to be projective.

DEFINITION 8.9. A functor $U : \mathfrak{C} \to \mathfrak{S}$ is called *strictly underlying* if
  (i) U is faithful.
  (ii) U has a left adjoint.
In this case the left adjoint $\mathrm{Fr} : \mathfrak{S} \to \mathfrak{C}$ is called a *free functor*.

Now suppose that $F : \mathfrak{C} \to \mathfrak{D}$, $G : \mathfrak{D} \to \mathfrak{C}$, and $F \overset{\eta}{\dashv} G$.

THEOREM 8.12. *If* G *preserves epics, then* F *preserves projectives.*

PROOF.   Suppose that $P \in \mathfrak{C}$ is projective and consider the diagram in $\mathfrak{D}$:

$$
\begin{array}{c}
FP \\
\downarrow \phi \\
X \xrightarrow{\;\;\epsilon\;\;\!\!\!\!\!\!\gg} Y
\end{array}
\tag{8.8}
$$

Applying η we get a diagram in $\mathfrak{C}$:

$$
\begin{array}{c}
P \\
\downarrow \eta(\phi) \\
GX \xrightarrow{\;\;G\epsilon\;\;\!\!\!\!\!\!\gg} GY
\end{array}
\tag{8.9}
$$

Here $G\epsilon$ is epic since G preserves epics. Since P is projective we can complete the diagram (8.9) by means of $\psi : P \to GX$ with $(G\epsilon)\psi = \eta(\phi)$, and then use η^{-1} to transfer to (8.8), thus proving that FP is projective.

Before proceeding, we enunciate the dual of Theorem 8.12; we again refer to an adjunction $F \dashv G$.

THEOREM 8.12*. *If* F *preserves monics,* G *preserves injectives.*

Now suppose that we are in a situation in which the underlying functor to sets $U : \mathfrak{C} \to \mathfrak{S}$ preserves epics; this is the case, for example, if $\mathfrak{C} = \mathfrak{G}$ or $\mathfrak{C} = \mathfrak{Ab}$. We may thus apply Theorem 8.12 to infer that $\mathrm{Fr} : \mathfrak{S} \to \mathfrak{C}$ preserves projectives. But since, as we have remarked, every object of $\mathfrak{S}$ is projective, it follows that *every free object of* $\mathfrak{C}$ *is projective.*

So far we have made no use of the requirement that an underlying functor to sets be faithful. The force of this requirement, as we will see, is that the rear of the adjunction $\mathrm{Fr} \dashv U$ is then epic.

THEOREM 8.13. *Let* $U : \mathfrak{C} \to \mathfrak{S}$ *be a strictly underlying functor to sets with left adjoint* Fr, *and let* $\delta : \mathrm{Fr} \circ U \to 1$ *be the rear of the adjunction. Then, for each* A *in* $\mathfrak{C}$,

$$
\delta_A : \mathrm{Fr}(UA) \to A
$$

is epic. Thus every object of $\mathfrak{C}$ *is, canonically, the epimorphic image of a free object.*

PROOF.   Let $\alpha, \beta : A \to B$ and $\alpha\delta_A = \beta\delta_A$. We apply the adjugant η to infer that $U(\alpha) = U(\beta)$. But U is faithful, so that $\alpha = \beta$.

If, also, U preserves epics, then δ_A is surjective (strictly, $U(\delta_A)$ is surjective), and $\mathrm{Fr}(UA)$ is projective, so that A is the surjective image of a projective object. Note also that if A is itself projective then δ_A has a right inverse $\nu:A \to \mathrm{Fr}(UA), \delta_A \nu = 1$. Let us call A a *retract* of B if there exist morphisms $\kappa:B \to A, \nu:A \to B$ with $\kappa\nu = 1$. Then we have

THEOREM 8.14. *Let $\mathfrak{C}$ be a category with a strictly underlying functor* U *to sets. Then every projective in $\mathfrak{C}$ is a retract of a free object. If, also,* U *preserves epics, then an object* A *in $\mathfrak{C}$ is projective if and only if it is a retract of a free object.*

PROOF. The first part is already proved. Since, under the hypothesis that U preserves epics, every free object of $\mathfrak{C}$ is projective, the proof of the theorem may be completed by demonstrating

PROPOSITION 8.15. *A retract of a projective is projective.*

PROOF. Suppose that we are given $\kappa:B \to A, \nu:A \to B$ with $\kappa\nu = 1$ and B projective. Let $\epsilon:X \to Y$ be epic, and let $\phi:A \to Y$ be given. Since B is projective, we may find $\psi:B \to X$ with $\epsilon\psi = \phi\kappa$. Then $\epsilon\psi\nu = \phi\kappa\nu = \phi$, so that A is projective.

Theorem 8.14 may be applied to the categories $\mathfrak{G}$, $\mathfrak{Ab}$ and leads to the conclusion, in these categories, that the projective objects are just the free objects; for a retract may be regarded as a subobject (provided that monics are injective) and subgroups of free groups (free abelian groups) are free (free abelian). In the categories of modules of Chapter 4, Theorem 8.14 provides a characterization of projectives.

EXERCISES

1.1.   Prove that axiom A3 implies that the identities 1_A in a category are unique.

1.2.   Prove that if f,g are invertible morphisms and fg is defined, then fg is invertible and

$$(fg)^{-1} = g^{-1}f^{-1}.$$

1.3.   Let $\mathfrak{C}_S$ be the category associated with a preordered set. Show that the equivalence classes of objects of $\mathfrak{C}_S$ form a new category in an obvious way.

1.4.   Suppose that, in a category $\mathfrak{C}$, an equivalence relation $\sim$ is defined on each set $\mathfrak{C}(X,Y)$ such that $fg \sim f'g'$ if $f \sim f'$, $g \sim g'$. Show how to form a *quotient* category $\mathfrak{C}/\sim$, in which the objects are the objects of $\mathfrak{C}$ and the morphisms are the equivalence classes of morphisms of $\mathfrak{C}$. Show that this generalizes the process of forming the quotient group of G by a normal subgroup H (recall the construction of the category $\mathfrak{C}_G$).

1.5. Show that the property expressed by Proposition 1.1 uniquely determines the zero morphisms. Using this property as *definition*, construct an example of a category $\mathfrak{C}$ with zero morphisms but without zero objects.

1.6. Offer a reasonable definition of the product $\mathfrak{C}_1 \times \mathfrak{C}_2$ of two categories.

2.1. Establish the facts cited in the first two examples of functors.

2.2. Generalize the differentiation functor to functions of several variables.

2.3. Show that
   (a) $U : \mathfrak{G} \to \mathfrak{S}$ is faithful but not full.
   (b) The embedding factor $\mathfrak{Ab} \subseteq \mathfrak{G}$ is full and faithful (this denotes the functor that acts as the identity on objects and morphisms).
   (c) Abel: $\mathfrak{G} \to \mathfrak{Ab}$ is not faithful.
   (d) Abel $\mathfrak{G} \to \mathfrak{Ab}$ is not full. (*Hint*: take $G_2 =$ quaternion group, $G_1 = \mathrm{Abel}(G_2) = G_2 / G_2'$.)

2.4. Let $Z(G)$ be the center of the group G. Show that $Z : \mathfrak{G} \to \mathfrak{G}$ is not, in any obvious way, a functor. Let $\mathfrak{G}_{\mathrm{epi}}$ be the subcategory of $\mathfrak{G}$ consisting of *surjective* homomorphisms. Show that $Z : \mathfrak{G}_{\mathrm{epi}} \to \mathfrak{G}$ is a functor. Is the image of Z contained in $\mathfrak{G}_{\mathrm{epi}}$?

2.5. A functor $\mathfrak{C}_1 \times \mathfrak{C}_2 \to \mathfrak{D}$ (see Exercise 1.6) is often called a *bifunctor*. Show that a bifunctor $F : \mathfrak{C}_1 \times \mathfrak{C}_2 \to \mathfrak{D}$ determines, for each $Y \in \mathfrak{C}_2$, a functor $F_Y : \mathfrak{C}_1 \to \mathfrak{D}$, and for each $X \in \mathfrak{C}_1$, a functor $F_X : \mathfrak{C}_2 \to \mathfrak{D}$, such that, for $f : X \to X'$ in $\mathfrak{C}$, $g : Y \to Y'$ in $\mathfrak{D}$,

$$F_{X'}(g)F_Y(f) = F_{Y'}(f)F_X(g).$$

Show, conversely, that, given functors

$$\bar{F}_Y : \mathfrak{C}_1 \to \mathfrak{D} \qquad \text{for each } Y \in \mathfrak{C}_2,$$

$$\bar{F}_X : \mathfrak{C}_2 \to \mathfrak{D} \qquad \text{for each } X \in \mathfrak{C}_1,$$

such that $\bar{F}_{X'}(g)\bar{F}_Y(f) = \bar{F}_{Y'}(f)\bar{F}_X(g)$, there is a unique bifunctor $F : \mathfrak{C}_1 \times \mathfrak{C}_2 \to \mathfrak{D}$ such that $F_X = \bar{F}_X, F_Y = \bar{F}_Y$.

2.6. Show how to embed a category $\mathfrak{C}$ with zero morphisms (see Exercise 1.5) in a category $\overline{\mathfrak{C}}$ with zero object. What can then be done to extend functors from $\mathfrak{C}$ to $\overline{\mathfrak{C}}$?

2.7. Show that $\mathfrak{C}(X, -)$ is a functor from $\mathfrak{C}$ to $\mathfrak{S}$, for all X in $\mathfrak{C}$. (The reader should interpret the symbol $\mathfrak{C}(X, -)$ appropriately!)

3.1. Show that (as asserted at the end of the section) τE and $H\tau$ are natural transformations and that $(H\tau)E = H(\tau E)$. Show also that $(\tau E)E' = \tau(EE')$, $H'(H\tau) = (H'H)\tau$, and that $\tau E, H\tau$ are equivalences if τ is an equivalence.

3.2. Let $\mathfrak{C},\mathfrak{D}$ be groups regarded as categories. What is a natural transformation of functors from $\mathfrak{C}$ to $\mathfrak{D}$? What is a natural equivalence of such functors?

3.3. Carry out an exercise similar to Exercise 3.2 for preordered sets regarded as categories.

3.4. (a) Let $A \in \mathfrak{C}$. Show that $\mathfrak{C}(A,-)$ is a functor from $\mathfrak{C}$ to $\mathfrak{S}$.

(b) (*The Yoneda lemma*) Let $F : \mathfrak{C} \to \mathfrak{S}$ be a functor. Show that the collection of natural transformations from $\mathfrak{C}(A,-)$ to F is in one-one correspondence with FA. (*Hint*: consider $1_A \in \mathfrak{C}(A,A)$.)

(c) Deduce that the collection of natural transformations from $\mathfrak{C}(A_2,-)$ to $\mathfrak{C}(A_1,-)$ is in one-one correspondence with $\mathfrak{C}(A_1,A_2)$. Show that the correspondence

$$\tau \Leftrightarrow f_\tau \qquad (f_\tau : A_1 \to A_2)$$

can be chosen so that

$$\sigma\tau \Leftrightarrow f_\tau f_\sigma.$$

Deduce that τ is a natural equivalence if and only if f_τ is a unit.

3.6. Let I,J be small categories. Establish natural equivalences

$$(\mathfrak{C}^I)^J \cong \mathfrak{C}^{I \times J} \cong (\mathfrak{C}^J)^I.$$

4.1. Identify the monomorphisms, epimorphisms, kernels, and cokernels in $\mathfrak{Ab}$ and $\mathfrak{S}_0$.

4.2. Show that the monomorphisms of $\mathfrak{G}$ are precisely the injections. Show that every surjection in $\mathfrak{G}$ is an epimorphism. Show that if $\phi : H \to G$ in $\mathfrak{G}$ is such that $\widehat{\phi H} \neq G$, then ϕ is not an epimorphism in $\mathfrak{G}$.

4.3. (This difficult exercise is designed to show that an epimorphism in $\mathfrak{G}$ is surjective.) Let H be a proper subgroup of G and let $\mathbf{Z}_2$ be written multiplicatively with generator η. Let G operate on the group $\mathbf{Z}_2^{G/H}$ (see Example 4, Section 2, of a functor) by

$$(g \cdot f)(Hg') = f(Hg'g).$$

Let $u \in \mathbf{Z}_2^{G/H}$ be given by $u(H) = \eta, u(Hg) = 1, g \notin H$. Show that $g \cdot u = u$ if and only if $g \in H$.

Form the semidirect product (Section 7 of Chapter 1) $E = \mathbf{Z}_2^{G/H} \times_\omega G$ for the given action of G on $\mathbf{Z}_2^{G/H}$ and define $\alpha,\beta : G \to E$ by $\alpha(g) = (1,g)$, $\beta(g) = (u(g \cdot u)^{-1},g)$. Show that α,β are homomorphisms such that $\alpha|H = \beta|H, \alpha \neq \beta$.

Deduce that if a homomorphism ϕ in $\mathfrak{G}$ is nonsurjective, it is not epic.

4.4. The *equalizer* of two morphisms f and g is a morphism u yielding a commutative diagram

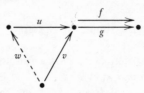

with w unique. Identify the equalizer in $\mathfrak{S}$, $\mathfrak{G}$, and $\mathfrak{Ab}$. Dualize the concept to *coequalizer* and identify the coequalizer in $\mathfrak{S}$ and $\mathfrak{Ab}$.

4.5. Explain how to compose two contravariant functors to produce a covariant functor. Similarly, compose a contravariant functor with a covariant functor.

4.6. Show precisely how the contravariant power set functor can be regarded as a functor to Boolean algebras.

4.7. Show that $\mathfrak{C}(-,X)$ is a contravariant functor from $\mathfrak{C}$ to $\mathfrak{S}$ (compare Exercise 2.7).

5.1. Identify the coproduct in $\mathfrak{S}, \mathfrak{G}, \mathfrak{Ab}, \mathfrak{S}_0$.

5.2. Let $\mathfrak{C}$ be a category with zero. Show that the following are equivalent:
   (a) X is a zero object.
   (b) $(X; 1, 1)$ is the product of X and X.
   (c) $(X; 1, 1)$ is the coproduct of X and X.

5.3. Let $\mathfrak{C}$ be a category, A_1, A_2 objects of $\mathfrak{C}$. Let $\mathfrak{C}/(A_1, A_2)$ be the category whose *objects* are pairs of morphisms (f_1, f_2), where $f_i : X \to A_i, i = 1, 2$; and in which a morphism $u : (f_1, f_2) \to (g_1, g_2)$, where $g_i : Y \to A_i$, is a morphism $u : X \to Y$ in $\mathfrak{C}$ with $f_i = g_i u$, $i = 1, 2$. Show that the product of A_1 and A_2 in $\mathfrak{C}$ is just a terminal object in $\mathfrak{C}/(A_1, A_2)$.

5.4. Suppose that, in $\mathfrak{C}$, every pair of objects has a product. Show how the product may be made into a functor.

5.5. Show that, in the category of cyclic groups, $\mathbf{Z}_m$ and $\mathbf{Z}_n$ have a product $\Leftrightarrow \mathbf{Z}_m$ and $\mathbf{Z}_n$ have a coproduct $\Leftrightarrow m$ and n are mutually prime.

5.6. Establish the fact that no infinite set of nonzero objects in the category of finitely generated abelian groups has a product. Can such a set have a coproduct?

5.7. Let Comm: $\mathfrak{G} \to \mathfrak{G}$ be the *commutator subgroup* functor. Show that Comm: $\mathfrak{G} \to \mathfrak{G}$, Abel: $\mathfrak{G} \to \mathfrak{G}$ and $\overline{\text{Abel}}$: $\mathfrak{G} \to \mathfrak{Ab}$ are product-preserving, but that, of these, only $\overline{\text{Abel}}$: $\mathfrak{G} \to \mathfrak{Ab}$ is coproduct-preserving.

5.8. Show that the underlying functors described in Section 2 are product-preserving, and that the free functors are coproduct-preserving.

6.1. Consider

(a) Show that if I and II are pullbacks, so is their composite.
(b) Dualize the above.
(c) Show that if the composite is a pullback and b is monic, then I is a pullback.
(d) Dualize the above.

6.2. Let

$$
\begin{array}{ccc}
\cdot & \xrightarrow{\psi_1} & \cdot \\
{\scriptstyle\psi_2}\downarrow & & \downarrow{\scriptstyle\phi_1} \\
\cdot & \xrightarrow[\phi_2]{} & \cdot
\end{array}
$$

be a pullback in $\mathfrak{G}$ (or $\mathfrak{Ab}$). Show that ψ_1 is epic if ϕ_2 is epic. Show that the "dual" also holds in $\mathfrak{Ab}$.

6.3. Let the diagram of Exercise 6.1 be in $\mathfrak{Ab}$. Show that if b,d are epic, a the kernel of b,c the kernel of d, and if f,h are isomorphisms, so is g. Hence show that if b,d are epic, a the kernel of b,c the kernel of d, and if h is an isomorphism, then I is a pushout.

6.4. "Dualize," in $\mathfrak{Ab}$, the final statement of Exercise 6.3.

6.5. Show that, if $\mathfrak{C}$ possesses finite products, then it possesses pullbacks if and only if it possesses equalizers (Exercises 4.4).

7.1. Let $\eta : \mathfrak{D}(FA,B) \cong \mathfrak{C}(A,GB)$ be the adjugant equivalence of an adjunction $F \dashv G$. Show that (7.2) precisely expresses the fact that η is a natural equivalence of functors $\mathfrak{C}^{op} \times \mathfrak{D} \to \mathfrak{S}$.

7.2. Prove Theorem 7.1.

7.3. Show how the equalizer (Exercise 4.2) may be regarded as right adjoint to a suitable constant functor.

7.4. Show how the coproduct may be regarded as left adjoint to a suitable constant functor.

7.5. Prove that a functor possessing a left adjoint preserves products, kernels, equalizers, and monics (Theorem 7.3).

7.6. Check that F', constructed in the proof of Theorem 7.7, is a functor.

8.1. Show that the (obvious) generalization of Theorem 4.1 of Chapter 2 is valid in any abelian category.

8.2. Show that the category of abelian p-groups, where p is prime, is an abelian category.

8.3. Show that Propositions 6.3, 6.4, and 6.4* remain valid in any abelian category.

8.4. Pursuing the connection between the square

and the sequence

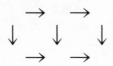

call the square *exact* if the sequence is exact. Show that, in the diagram below, the composite square is exact if the individual squares are exact (i) in $\mathfrak{Ab}$, (ii) in any abelian category,

8.5. Show that, in $\mathfrak{Ab}$, the square

$$\begin{array}{ccc} & \xrightarrow{\psi_1} & \\ \psi_2\downarrow & & \downarrow\phi_1 \\ & \xrightarrow{\phi_2} & \end{array}$$

is exact if and only if $\psi_1\psi_2^{-1}=\phi_1^{-1}\phi_2$.

8.6. Give an example of an additive functor that is neither left nor right exact. Give an example of a nonadditive functor $\mathfrak{Ab} \to \mathfrak{Ab}$.

8.7. (Harder) Let $\mathfrak{A}$ be an abelian category and I a small category. Show that $\mathfrak{A}^I$ is abelian. (*Hint*: the kernel of a diagram of morphisms is the diagram of kernels.)

4

MODULES

1. RINGS

A *ring* is a triple $(R; +, *)$, consisting of a set R and two binary operations, addition $(+)$ and multiplication $(*)$, subject to the axioms:

$\mathbf{R}_1$: R is an abelian group under $+$.

$\mathbf{R}_2$: Multiplication is associative.

$\mathbf{R}_3$: The distributive laws

$$r * (s + t) = r * s + r * t,$$

$$(r + s) * t = r * t + s * t, \quad r, s, t \epsilon R$$

hold.

$\mathbf{R}_4$: There is an *identity* element $1 \neq 0$, such that $1 * r = r * 1 = r, r \epsilon R$.

Note. The customary definition of a ring omits axiom $\mathbf{R}_4$. Then a ring satisfying $\mathbf{R}_4$ is called a *unitary* ring. Since the only rings considered in this book are unitary, we have preferred to incorporate $\mathbf{R}_4$ into the definition of a ring, so that we may omit the word "unitary."

We will usually write multiplication merely as juxtaposition; and we will (as for other structures) refer to R itself as a ring. The ring R is said to be *commutative* if multiplication is commutative; it is an *integral domain* if it is commutative and if $rs \Rightarrow 0 \Rightarrow r = 0$ or $s = 0$, $r, s \in R$; it is a *division ring* if $R - (0)$ is a group under multiplication; and a *field* if it is a commutative division ring.

EXAMPLES.

(1) $(\mathbf{Z}; +, \times)$ is an integral domain.

(2) $(\mathbf{Z}_n; +, \times)$ is a commutative ring, where $+, \times$ are induced by the operations in $\mathbf{Z}$. The reader should prove that $\mathbf{Z}_n$ is a field if and only if n is prime.

(3) The following is an important trick that we will exploit in the sequel.

If $(R; +, *)$ is a ring, then $(R^0; +, *^0)$ is a ring, where $R^0 = R$ (as sets) and

$$r *^0 s = s * r.$$

We call R^0 the *opposite* ring of R. Of course, $R = R^0$ if and only if R is commutative, and $R^{00} = R$.

(4) If A is an abelian group, then $\text{End}(A) = \text{Hom}(A, A)$, the set of *endomorphisms* of A, is a ring, where $(f + g)(a) = f(a) + g(a)$, $(f * g)(a) = f(g(a))$, $f, g \in \text{End}(A), a \in A$. The reader should check the axioms.

(5) Let G be a group and R a ring. We form the set of formal linear combinations

$$\sum_{\sigma \in G} r_\sigma \sigma, \quad r_\sigma \in R,$$

where, of course, $r_\sigma = 0$ except for a finite number of elements σ of G. We add by the rule

$$\sum_\sigma r_\sigma \sigma + \sum_\sigma s_\sigma \sigma = \sum_\sigma (r_\sigma + s_\sigma) \sigma,$$

and we multiply by the rule

$$\left(\sum_\sigma r_\sigma \sigma \right) \left(\sum_\sigma s_\sigma \sigma \right) = \sum_{\sigma, \tau} r_\sigma s_\tau \sigma \tau.$$

In this way we obtain a ring $R[G]$, called the *group ring* of G over R; we leave the reader to check the axioms. Notice that $R[G]$ is commutative if R and G are commutative.

(6) If R is a ring, then we may form the ring of *polynomials* $R[x]$ in the *indeterminate* x, with *coefficients* in R, in the usual way. We may also form the ring of polynomials in any number of variables. We may also form the ring of *formal power series* $R[[x]]$, where the rules for addition and multiplication are exactly as for polynomials. It is not hard to prove that $R[x]$ is commutative if (and only if) R is commutative, and an integral domain if (and only if) R is an integral domain.

We list the following elementary properties of rings, leaving the proofs as an exercise.

PROPOSITION 1.1. *Let* R *be a ring. Then*

(*i*) $0r = r0 = 0$.

(*ii*) $(-r)s = r(-s) = -rs$.

To form a *category* of rings, we must specify the morphisms; thus a function $f: R \to S$ from the ring R to the ring S is a *ring homomorphism* if it preserves addition, multiplication, and identity element, that is, if

$$f(r_1 + r_2) = fr_1 + fr_2, \quad f(r_1 r_2) = (fr_1)(fr_2), \quad f1 = 1.$$

We will often abbreviate "ring homomorphism" to "homomorphism," especially where the context makes it clear that a map of rings is in question. Clearly, ring homomorphisms inherit certain properties from homomorphisms of abelian groups. We study this question in the following propositions.

PROPOSITION 1.2. *A ring homomorphism $\phi: R \rightarrow S$ is an isomorphism (invertible) if and only if it is surjective and injective.*

PROOF. Of course, an isomorphism is surjective and injective. Conversely, let $\phi: R \rightarrow S$ be surjective and injective. Then ϕ has an inverse $\phi^{-1}: S \rightarrow R$ which we know to be a homomorphism of abelian groups. Thus it only remains to show that ϕ^{-1} preserves multiplication and identity element. Now if $s_1, s_2 \in S$, let $\phi r_1 = s_1, \phi r_2 = s_2$. Then $\phi(r_1 r_2) = s_1 s_2$, so that $\phi^{-1}(s_1 s_2) = r_1 r_2 = \phi^{-1}(s_1) \phi^{-1}(s_2)$. Also $\phi(1) = 1$, so that $\phi^{-1}(1) = 1$.

PROPOSITION 1.3. *A ring homomorphism $\phi: R \rightarrow S$ is injective if and only if $\phi^{-1}(0) = 0$.*

PROOF. Since the assertion is true of abelian group homomorphisms it must be true of ring homomorphisms.

Let R be a ring and let I be a subgroup of the abelian group R. Then I is said to be

a *left ideal* if $r \in R, a \in I \Rightarrow ra \in I$,

a *right ideal* if $r \in R, a \in I \Rightarrow ar \in I$,

a *two-sided ideal* if $r \in R, a \in I \Rightarrow ra \in I, ar \in I$.

We define the *kernel* of $\phi: R \rightarrow S$ to be the kernel of ϕ as group homomorphism.

THEOREM 1.4. *The kernel of a ring homomorphism $\phi: R \rightarrow S$ is a two-sided ideal in* R. *Conversely, given any two-sided ideal* $I \neq R$ *in* R, *there is a quotient ring* R/I, *and a projection homomorphism* $\pi: R \rightarrow R/I$ *with kernel* I.

PROOF. The kernel of ϕ is certainly an abelian subgroup of R. Moreover if $r \in R, a \in \ker \phi$, then

$$\phi(ra) = (\phi r)(\phi a) = (\phi r)0 = 0, \quad \phi(ar) = (\phi a)(\phi r) = 0(\phi r) = 0,$$

by Proposition 1.1(i). Thus $\ker \phi$ is a two-sided ideal.

Conversely, form the abelian quotient group R/I. We must introduce a multiplication into R/I so that R/I becomes a ring and $\pi: R \rightarrow R/I$ becomes a ring homomorphism. The unique multiplication which would satisfy the second requirement is given by the rule

$$(r_1 + I)(r_2 + I) = r_1 r_2 + I. \tag{1.1}$$

To see that (1.1) is a well-defined operation on R/I, we observe that, if $a_1, a_2 \in I$,

$$(r_1 + a_1)(r_2 + a_2) = r_1 r_2 + (a_1 r_2 + r_1 a_2 + a_1 a_2) \equiv r_1 r_2 \bmod I,$$

since I is a two-sided ideal. Plainly the multiplication (1.1) satisfies $\mathbf{R}_2, \mathbf{R}_3, \mathbf{R}_4$ (with the identity element $1 + I$), and π is then a ring homomorphism with kernel I. Notice that we must take $I \neq R$, so that $1 \neq 0$ in R/I.

We leave the reader to prove

PROPOSITION 1.5. *If $\phi: R \to S$ is a ring homomorphism, then ϕR is a subring of S, isomorphic to $R/\ker \phi$.*

Notice that, in the strict categorical sense, the category of rings and homomorphisms does not have kernels and cokernels. We take the abelian group kernel, but this is never a *unitary* ring, because 1 is never in the kernel. Likewise we may take the abelian group cokernel but this also fails to be a unitary ring. In general, it is not even a nonunitary ring, since ϕR fails to be an ideal. Indeed, ϕR is only an ideal when $\phi R = S$!

Henceforth (and in the last sentence!) we write "ideal" for "two-sided ideal." We now turn our attention to elementary properties of ideals.

We leave the reader to prove

PROPOSITION 1.6. *The intersection of any family of (left, right, two-sided) ideals of R is a (left, right, two-sided) ideal of R.*

We denote by $(a_1, \ldots, a_n)$ the ideal *generated* by $a_1, \ldots, a_n$; that is, the smallest ideal of R containing $a_1, \ldots, a_n$. If $n = 1$, we call the ideal *principal*. Plainly we may also consider nonfinitely generated ideals; and, furthermore, it is clear that the concepts above may be applied to left or right ideals. Indeed the left ideal generated by $a_1, \ldots, a_n$ consists precisely of the elements

$$r_1 a_1 + \cdots + r_n a_n, \quad r_i \in R, \quad i = 1, \ldots, n. \tag{1.2}$$

An ideal M in R is called *maximal* if $M \neq R$ and M is not contained in any other ideal except R itself. [Again, we may talk of maximal left (right) ideals.] An ideal P of a commutative ring is called *prime* if, whenever $rs \in P, r \in P$ or $s \in P$. It is customary to exclude 0 and R as prime ideals; we will adhere to this custom. We do not apply the concept prime ideal to noncommutative rings.

Note that, if R is commutative, then $b|a$ ("b divides a") if and only if $a \in (b)$, the ideal generated by b.

THEOREM 1.7. *If R is commutative, every nonzero maximal ideal is prime.*

PROOF.   Suppose that M is maximal and $ab \in M$. We must prove that $a \in M$ or $b \in M$. It suffices to suppose that $a \notin M$ and deduce that $b \in M$. Let I be the ideal generated by M and a. Since $a \notin M, I$ properly contains M. But M is maximal, so that $I = R$. Thus we may write the identity of R as

$$1 = m + ra, \quad m \in M, \quad r \in R.$$

Then $b = mb + rab, mb \in M, rab \in M$, so that $b \in M$.

That the converse is, in general, false may be seen by the example $R = \mathbf{Z}[x], I = (x)$. Then I is certainly prime since if p, q are two polynomials whose product is divisible by x, either p or q must be divisible by x. On the other hand, (x) is not maximal, since, for example, $(x) \subset (2, x) \neq \mathbf{Z}[x]$. We will see later that the converse of Theorem 1.7 holds in a *principal ideal domain*, that is, an integral domain in which every ideal is principal.

THEOREM 1.8.   *If* R *is commutative, then the ideal* I *is maximal if and only if* R/I *is a field.*

PROOF.   Let I be maximal. The proof that R/I is a field is very much like the proof of Theorem 1.7. Let $x + I$ be a nonzero element of R/I. Then $x \notin I$, so that the ideal generated by I and x is R. Thus 1 may be written as $1 = a + xy, a \in I, y \in R$. But then $(x + I)(y + I) = xy + I = 1 + I$, so that $x + I$ has an inverse and R/I is a field. We leave the reader to prove the converse.

THEOREM 1.9.   *If* R *is commutative, then the ideal* I *is prime if and only if* R/I *is an integral domain.*

PROOF.   Let I be prime and suppose that $(x + I)(y + I) = 0 \in R/I$. But $(x + I)(y + I) = xy + I$, so the hypothesis means that $xy \in I$. Since I is prime, $x \in I$ or $y \in I$, so $x + I = 0$ or $y + I = 0$, showing that R/I is an integral domain. We again leave the converse to the reader.

Of course, Theorems 1.8 and 1.9 provide another (and, in some respects, better) proof of Theorem 1.7.

THEOREM 1.10.   *The ring* R *possesses maximal (left, right, two-sided) ideals.*

PROOF.   We will be content to consider left ideals. We use Zorn's lemma. Consider the set $\mathfrak{F}$ of left ideals of R disjoint from 1. This set is nonempty, since $(0) \in \mathfrak{F}$, and it is inductively ordered by inclusion. For if $\{I_\alpha\}$ is a totally ordered set of left ideals of R disjoint from 1, then $\cup_\alpha I_\alpha$ is a left ideal of R disjoint from 1 and $I_\alpha \subseteq \cup_\alpha I_\alpha$. Thus $\mathfrak{F}$ has a maximal element, and this maximal element is, of course, a maximal left ideal.

We now give some examples of ideals.

EXAMPLES.

1. (0) and R are ideals in R. These are called *improper* ideals, other ideals being proper. Of course, every unit generates the improper ideal R. (The definition of a *unit* immediately follows this set of examples.)

2. In $\mathbf{Z}$ the proper ideals are precisely the subgroups $n\mathbf{Z}, n \geqslant 2$. The maximal ideals are the prime ideals, and these are the ideals $p\mathbf{Z}$, with p prime.

3. A division ring has no proper ideals.

4. The proper ideals of $\mathbf{Z}_k$ are the sets $l\mathbf{Z}_k$, where $l | k, l \neq 1, k$.

5. In the ring of $(n \times n)$-matrices over a ring R, the set of *k*th *row matrices*, I_k, consisting of matrices $(a_{ij} | a_{ij} \in R, a_{ij} = 0, i \neq k)$, is a right ideal. Similarly the set of kth column matrices is a left ideal.

6. In the ring of $(n \times n)$-matrices over $\mathbf{Z}$, the set of matrices (m_{ij}) such that m_{1j} is even, $1 \leqslant j \leqslant n$, is a right ideal. Similarly the set of matrices (m_{ij}) such that m_{j1} is even, $1 \leqslant i \leqslant n$, is a left ideal. The set of matrices with *all* entries even is a two-sided ideal.

We say that $u \in R$ is a *unit* of R if there exists $v \in R$ with $uv = vu = 1$. (We write $v = u^{-1}$ and call it the *inverse* of u; it is plainly uniquely determined by u.) This is equivalent to asking that both the left ideal and the right ideal generated by u be R itself. For, if the latter condition holds we have elements v, w such that $vu = 1, uw = 1$; but then $v = vuw = w$. It is plain that the units of R form a multiplicative group, $\mathrm{Un}(R)$, and Un is then a functor

$$\mathrm{Un}: \mathfrak{R}_1 \to \mathfrak{G}$$

from the category of rings to the category of groups, since a homomorphism $R \to S$ must send the units of R to units of S.

We now discuss again Example 4 of Section 7 in Chapter 3.

THEOREM 1.11. *Let* $\mathbf{Z}[G]$ *be the integral group ring of the group* G *(see Example 5 of a ring). Then* $\mathbf{Z}[-]$ *is a functor* $\mathfrak{G} \to \mathfrak{R}_1$, *left adjoint to* $Un: \mathfrak{R}_1 \to \mathfrak{G}$.

PROOF. We must establish a natural equivalence

$$\eta: \mathrm{Hom}\,(\mathbf{Z}[\,G\,], R) \cong \mathrm{Hom}\,(\,G, \mathrm{Un}\,(R)).$$

Now plainly the elements of G, regarded as elements of $\mathbf{Z}[G]$ by means of the identification $\sigma \leftrightarrow 1\sigma$, are units of $\mathbf{Z}[G]$, retaining their inverses in G, so that $G \subseteq \mathrm{Un}\,\mathbf{Z}[G]$. It is thus plain that any ring homomorphism $\phi: \mathbf{Z}[G] \to R$ determines, by restriction, a homomorphism $\phi_0: G \to \mathrm{Un}(R)$ and we define

$$\eta(\phi) = \phi_0.$$

It is plain that η is natural.

Now let $\psi: G \to \mathrm{Un}(R)$ be a group homomorphism. It is plain from the definition of $\mathbf{Z}[G]$ that, ignoring the multiplication, it is a free abelian group on the set G as basis. Thus ψ, regarded as a function from G to the underlying abelian group of R, has a unique extension to an abelian group homomorphism $\bar{\psi}: \mathbf{Z}[G] \to R$. But $\bar{\psi}$ is then, in fact, a ring homomorphism. For $\bar{\psi}(1e) = \psi(e) = 1$ (where e is the identity of G) and

$$\bar{\psi}\Big(\big(\sum m_\sigma \sigma\big)\big(\sum n_\tau \tau\big)\Big) = \bar{\psi}\Big(\sum m_\sigma n_\tau \sigma\tau\Big) = \sum m_\sigma n_\tau \psi(\sigma\tau)$$

$$= \sum m_\sigma n_\tau \psi(\sigma)\psi(\tau), \text{ because } \psi \text{ is a homomorphism}$$

$$= \big(\sum m_\sigma \psi(\sigma)\big)\big(\sum n_\tau \psi(\tau)\big)$$

$$= \bar{\psi}\big(\sum m_\sigma \sigma\big)\bar{\psi}\big(\sum n_\tau \tau\big).$$

We define

$$\bar{\eta}(\psi) = \bar{\psi}.$$

Plainly, $\eta\bar{\eta}(\psi) = \psi$. To check that $\bar{\eta}\eta(\phi) = \phi$, it is sufficient to check that this relation holds on the generating set G of the free abelian group $\mathbf{Z}[G]$; but, on G, it is obvious. Thus η is a natural equivalence.

We say that two elements a,b of R are *right associated* if there exists a unit u with $a = bu$. Plainly the relation of being right associated is an equivalence relation. Alternatively, a,b are right associated if they generate the same right ideal. There is, of course, a similar notion of *left associated*, and any element right or left associated to a unit is itself a unit.

2. MODULES

DEFINITION 2.1. Let R be a ring. Then a *left R-module* is a pair (A, ρ) consisting of abelian group A and a homomorphism $\rho: R \to \mathrm{End}\, A$. A *right R-module* is a left R^0-module.

We often speak of the left R-module A if ρ may be understood. This way of speaking is reinforced by the following important observation. Suppose that we write ra for $\rho(r)(a)$. Then the fact that $\rho(r)$ is an endomorphism of A is exactly reflected in the rule

$\mathbf{M}_1$: $r(a+b) = ra + rb, \quad a,b \in A, \quad r \in R,$

while the fact that ρ is a ring homomorphism is exactly reflected in the rules

$\mathbf{M_2}$: $(r_1 + r_2)a = r_1a + r_2a, \quad a \in A, \quad r_1, r_2 \in R,$

$\mathbf{M_3}$: $(r_1 r_2)a = r_1(r_2 a), \quad a \in A, r_1, r_2 \in R,$

$\mathbf{M_4}$: $1a = a, a \in A.$

Thus we may, equivalently, define a left R-module to be an abelian group A together with a pairing (operation) $R \times A \to A$, written $(r, a) \mapsto ra$, subject to the axions $\mathbf{M_1}$ to $\mathbf{M_4}$. We will use this way of writing, which, of course, explains the presence of the term "left" in the definition. A right R-module B leads to the notation $br, b \in B, r \in R$, with rules analogous to $\mathbf{M_1}$ to $\mathbf{M_4}$. Of course it is only in axiom $\mathbf{M_3}$ that the difference between a left R-module and a right R-module is essential. If R is commutative, there is no essential difference.

EXAMPLES.
   1. An abelian group has a natural structure of $\mathbf{Z}$-module, given by assigning to $na, n \in \mathbf{Z}, a \in A$, its standard meaning.
   2. If R is a field, an R-module is just a vector space over R.
   3. A left ideal in R is a left R-module, a right ideal is a right R-module.
   4. The set of $n \times n$-matrices over R becomes a left R-module when it is given the obvious (additive) abelian group structure and we define $r(a_{ij}) = (ra_{ij}), r, a_{ij} \in R$. Similarly, it may be given the structure of a right R-module.
   5. Consider the groupring $R[G]$ of G over R. Then R becomes a left $R[G]$-module under the rule $(\sum r_\sigma \sigma)r = \sum r_\sigma r$. Similarly it may be made into a right $R[G]$-module.
   6. Let A be a left S-module and $\phi: R \to S$ a ring homomorphism. Then we may form a left R-module A^ϕ whose underlying abelian group is that of A and whose R-module structure is given by $ra = \phi(r)a$. Similarly a right S-module B is converted into a right R-module B^ϕ.
   7. Let R be a ring, A an abelian group. Then the abelian group $R \otimes A$ acquires a left R-module structure by the rule $r(r' \otimes a) = rr' \otimes a, r, r' \in R, a \in A$. (Notice that it is sufficient, in the light of axiom $\mathbf{M_1}$, to describe the operation of R on *generators* of the abelian group in question.) Similarly, we give $A \otimes R$ a right R-module structure.

For the rest of this section we concentrate on *left* R-modules but, of course, all our remarks have analogs for right R-modules. We first make the collection of left R-modules into a category $_R\mathfrak{M}$ by the following definition.

DEFINITION 2.2. If A, B are left R-modules a *homomorphism* $\phi : A \to B$ is a homomorphism of the underlying abelian group structures such that

$$\phi(ra) = r\phi(a), \quad a \in A, \quad r \in R.$$

We saw in Section 1 of Chapter 2 how to give the set $\operatorname{Hom}(A, B)$ of homomorphisms from the abelian group A to the abelian group B the structure of an abelian group. It is then obvious that if A, B are left R-modules, the set $\operatorname{Hom}_R(A, B)$ of R-module homomorphisms from A to B is a subgroup of $\operatorname{Hom}(A, B)$. For $0 : A \to B$ is plainly an R-module homomorphism, and if $\phi, \psi : A \to B$ are R-module homomorphisms then

$$(\phi - \psi)(ra) = \phi(ra) - \psi(ra) = r\phi(a) - r\psi(a) = r(\phi(a) - \psi(a)) = r(\phi - \psi)(a).$$

It is also obvious that if A, B are abelian groups and we view them (as in Example 1) as $\mathbf{Z}$-modules, then $\operatorname{Hom}_{\mathbf{Z}}(A, B) = \operatorname{Hom}(A, B)$. Thus we continue to write $\operatorname{Hom}(A, B)$ for $\operatorname{Hom}_{\mathbf{Z}}(A, B)$. We denote the category of right R-modules by $\mathfrak{M}_R$. Then

$$_R\mathfrak{M} = \mathfrak{M}_{R^0} \tag{2.1}$$

under the canonical identification. We leave the reader to show that a morphism of $_R\mathfrak{M}$ is an isomorphism if and only if it is an isomorphism of abelian groups, that is, if and only if it is injective and surjective.

It is evident what we should understand by a *submodule*. Moreover, if A' is a submodule of A, the quotient abelian group $A'' = A/A'$ admits a unique R-module structure such that the projection $\pi : A \twoheadrightarrow A''$ is an R-module homomorphism; namely, we define

$$r(a + A') = ra + A'. \tag{2.2}$$

The reader will easily verify that (2.2) does define an R-module structure in A'', which is then called a *quotient module* of A.

THEOREM 2.1. *Let* $\phi : A \to B$ *in* $_R\mathfrak{M}$. *Then*

(i) $\phi^{-1}(0)$ *is a submodule of* A *and is the kernel of* ϕ *in* $_R\mathfrak{M}$.

(ii) ϕA *is a submodule of* B.

(iii) $B/\phi A$ *is the cokernel of* ϕ *in* $_R\mathfrak{M}$.

PROOF.

(i) We only have to observe that, if $\phi a = 0$, then $\phi(ra) = 0$. But $\phi(ra) = r\phi(a) = r0 = 0$. Since $\phi^{-1}(0)$ is a module, the embedding of $\phi^{-1}(0)$ in A is the categorical kernel.

(ii) We only have to observe that if $b \in \phi A$, then $rb \in \phi A$. But if $b = \phi a$, $rb = r\phi a = \phi(ra)$.

(iii) Since ϕA is a submodule of $B, B/\phi A$ acquires a module structure by

(2.2). Then, certainly, the projection $\pi : B \to B/\phi A$ is the categorical co-kernel.

The following proposition is trivial in view of what is known for abelian groups.

PROPOSITION 2.2

(*i*) *The monics in* $_R\mathfrak{M}$ *are precisely the injective homomorphisms. A homomorphism is injective if and only if its kernel is zero.*

(*ii*) *The epics in* $_R\mathfrak{M}$ *are precisely the surjective homomorphisms. A homomorphism is surjective if and only if its cokernel is zero.*

We also take from abelian group theory the notion of *exact sequences*. In particular we remark that every homomorphism $\phi : A \to B$ in $_R\mathfrak{M}$ gives rise naturally to an exact sequence

$$0 \to \ker\phi \xrightarrow{\iota} A \xrightarrow{\phi} B \xrightarrow{\pi} \mathrm{coker}\,\phi \to 0 \qquad (2.3)$$

Now a short exact sequence

$$0 \to A' \xrightarrow{\mu} A \xrightarrow{\epsilon} A'' \to 0 \qquad (2.4)$$

is characterized by the properties

(i) μ is monic.

(ii) ϵ is epic.

(iii) ker $\epsilon = $ im μ.Thus, effectively, (2.4) asserts that ϵ is an epimorphism from A to A'' with kernel A'. We prove

PROPOSITION 2.3. *The sequence* $A' \xrightarrow{\mu} A \xrightarrow{\epsilon} A''$ *is short exact if and only if* μ *is monic,* $\epsilon\mu = 0$, *and* ϵ *induces an isomorphism of* coker μ *onto* A''.

PROOF. Suppose that the sequence is short exact. Then certainly μ is monic and $\epsilon\mu = 0$. Then ϵ induces

$$\bar{\epsilon} : \mathrm{coker}\,\mu \to A''$$

which is certainly epic. However it is also monic (as demonstrated in the case of abelian groups). Thus $\bar{\epsilon}$ is an isomorphism.

Conversely if ϵ induces an isomorphism $\bar{\epsilon} : \mathrm{coker}\,\mu \cong A''$, then ϵ is certainly epic and ker$\epsilon = im\mu$; again, the abelian group argument holds without change.

Thus we may replace item (iii) in the characterization of a short exact sequence by

(*iii*)′ coker $\mu = $ im ϵ.

More generally the sequence $\ldots \to A_{n-1} \overset{\phi_{n-1}}{\to} A_n \overset{\phi_n}{\to} A_{n+1} \to \cdots$ is exact at A_n if $\operatorname{im}\phi_{n-1} = \ker\phi_n$ (this is the definition) or, equivalently, if $\operatorname{coker}\phi_{n-1} = \operatorname{im}\phi_n$. We may say (without giving precise definitions!) that

$$\text{a sequence is exact if and only if it is coexact.} \qquad (2.4)$$

THEOREM 2.4. **(Noether Isomorphism Theorems)**

(*i*) *If* A_1, A_2 *are submodules of some module* A *and* $A_1 + A_2$ *is their sum, that is, the smallest submodule of* A *containing* A_1 *and* A_2, *then*

$$A_1 + A_2 / A_1 \cong A_2 / A_1 \cap A_2.$$

(*ii*) *If* $A \subseteq B \subseteq C$ *in* $_R\mathfrak{M}$, *then* $A/C \big/ B/C \cong A/B$.

PROOF

(i) Plainly $A_1 + A_2$ is the submodule of A consisting of elements $a_1 + a_2$, $a_i \in A_i, i = 1, 2$. We consider the homomorphism $\phi = \pi\iota : A_2 \to A_1 + A_2 / A_1$,

$$A_2 \overset{\iota}{\rightarrowtail} A_1 + A_2 \overset{\pi}{\twoheadrightarrow} A_1 + A_2 / A_1.$$

Then ϕ is surjective. For, plainly, every element of $A_1 + A_2 / A_1$ may be written $a_2 + A_1$, and $a_2 + A_1 = \phi a_2$. Also $\ker\phi = A_1 \cap A_2$. For $\ker\pi = A_1$, so $\ker\phi = \ker\pi\iota$ consists of the elements of $\ker\pi$ in the domain A_2 of ϕ, that is, $\ker\phi = A_1 \cap A_2$.

Thus, ϕ induces an isomorphism $A_2 / A_1 \cap A_2 \cong A_1 + A_2 / A_1$.

(ii) The proof for groups (Theorem 4.11 of Chapter 1) goes through without change.

THEOREM 2.5. *Let* $A_i, i \in I$, *be a family of* R-modules. *Then their abelian group product* ΠA_i *may be given the structure of an* R-module *by the rule*

$$r\{a_i\} = \{ra_i\} \qquad (2.5)$$

in such a way that it becomes the product in $_R\mathfrak{M}$. *Indeed* (2.5) *is the unique module structure in* ΠA_i *making the projections* $\Pi A_i \to A_i$ R-module *homomorphisms.*

Dually, the abelian group coproduct $\oplus A_i$ *may be given the structure of an* R-module *by the same rule* (2.5) *in such a way that it becomes the coproduct in* $_R\mathfrak{M}$. *Indeed,* (2.5) *is the unique module structure in* $\oplus A_i$ *making the injections* $A_i \to \oplus A_i$ R-module *homomorphisms.*

Each step of the proof is trivial; we leave the details to the reader.

The product and coproduct given in Theorem 2.5 are also called *direct product* and *direct sum*; we will use these terms together with the cognate terms *direct factor* and *direct summand*, when convenient. We may restate the essence of Theorem 2.5 by saying that the underlying functor $U: {}_R\mathfrak{M} \to \mathfrak{Ab}$ preserves and reflects products and coproducts; we have already seen that U preserves and reflects exactness. It is thus clear why many propositions about modules are proved by exactly the arguments used for abelian groups. For example, using the functor U we may appeal to abelian group theory (in fact, Theorem 4.1 of Chapter 2) for a proof of the following theorem.

THEOREM 2.6. *Let* $0 \to A' \xrightarrow{\mu} A \xrightarrow{\epsilon} A'' \to 0$ *be a short exact sequence. Then the following statements are equivalent*:

(*i*) *There exists* $\eta: A \to A'$ *with* $\eta\mu = 1$.
(*ii*) *There exists* $\eta: A \to A'$ *with* $(A; \eta, \epsilon)$ *the product of* A' *and* A''.
(*iii*) *There exists* $\nu: A'' \to A$ *with* $\epsilon\nu = 1$.
(*iv*) *There exists* $\nu: A'' \to A$ *with* $(A; \mu, \nu)$ *the coproduct of* A' *and* A''.
(*v*) *There exist* $\eta: A \to A', \nu: A'' \to A$ *with* $\mu\eta + \nu\epsilon = 1$.

If any of these conditions hold we say that the sequence *splits*.

COROLLARY 2.7. *If* $\pi: A \to A$ *has the property that* $\pi^2 = \pi$ *then*

$$A \cong \mathrm{Im}\,\pi \oplus \ker \pi.$$

PROOF. The exact sequence

$$0 \to \ker \pi \xrightarrow{\mu} A \xrightarrow{\epsilon} \mathrm{Im}\,\pi \to 0,$$

where $\epsilon a = \pi a$, splits by $\nu: \mathrm{Im}\,\pi \to A$, given by $\nu(a) = a, a \in \mathrm{Im}\,\pi$. For if $a = \pi b$,

$$\epsilon\nu a = \pi(\pi b) = \pi b = a.$$

We say, as in Chapter 1, that an R-module A is the *internal direct sum* of its submodules A_i if $(A; \iota_i)$ is the coproduct of the modules A_i, where ι_i is the inclusion $A_i \subseteq A$. We may then write $A = \bigotimes A_i$ if we wish to be very precise. Then the conclusion of Corollary 2.7 is that $A = \mathrm{Im}\,\pi \bigotimes \ker \pi$; but it would not be misleading to write $A = \mathrm{Im}\,\pi \oplus \ker \pi$.

Just as for groups, we have the following theorem characterizing the internal direct sum; we write $\sum A_i$ for the smallest submodule of A containing each of the submodules A_i of A.

THEOREM 2.8. *Let* $A_i, i \in I$, *be a family of submodules of* A. *Then the following statements are equivalent*:

(*i*) $A = \bigotimes A_i$.
(*ii*) *Each element* $a \in A$ *is uniquely expressible as* $a = \sum a_i, a_i \in A_i$, *with almost all* a_i *zero*.

(*iii*) $A = \sum A_i$ and, for each $i, A_i \cap \sum_{j \neq i} A_j = 0$.

Given any R-module A and set S of elements of A, let A' be the smallest submodule of A containing S. We say that A' is *generated* by S. If $A' = A$, then S is said to be a set of generators of A, and A is *finitely generated (cyclic)* if it has a finite (singleton) set of generators. Plainly if $A = \sum A_i$ and S_i is a set of generators of A_i, then $\cup S_i$ is a set of generators of A.

THEOREM 2.9. *Let S be a set of elements of the module A, and let A' be the submodule of A generated by S. Then A' consists precisely of the elements of A expressible as finite sums*

$$\sum r_i a_i, \quad r_i \in R, \quad a_i \in S.$$

PROOF. Certainly any submodule of A containing S must contain such finite sums $\sum r_i a_i$. On the other hand, the collection of all such finite sums plainly forms a submodule of A and must therefore be the smallest submodule containing S.

COROLLARY 2.10. *Let $\phi : A \to B$ be a homomorphism of R-modules and let A be generated by the set S. Then ϕ is entirely determined by $\phi|S$.*

PROOF. Every element of A is expressible as a finite sum $\sum r_i a_i$, $r_i \in R, a_i \in S$. But

$$\phi\left(\sum r_i a_i \right) = \sum r_i \phi(a_i),$$

so that ϕ is determined by its values on the elements a_i of S.

We close this section by summing up a number of the assertions of this section in the following abstract form.

THEOREM 2.11. *The category $_R\mathfrak{M}$ is abelian.*

The reader should go carefully through the axioms of an abelian category, checking their validity in the case of $_R\mathfrak{M}$.

3. THE FUNCTOR HOM

In this section we study the functors $\text{Hom}_R(A, -): {}_R\mathfrak{M} \to \mathfrak{Ab}$, $\text{Hom}_R(-, B): {}_R\mathfrak{M}^0 \to \mathfrak{Ab}$ together with their analogs for right modules. However, it turns out that we obtain the most satisfactory theory by enriching the notion of a module.

DEFINITION 3.1. An (R, S)-*bimodule* is an abelian group A that is both a left R-module and a right S-module and satisfies the further compatibility (associativity) condition

$$(ra)s = r(as), \quad r \in R, \quad a \in A, \quad s \in S. \tag{3.1}$$

The entire theory of Section 2 could have been executed for (R,S)-bimodules, at the mere cost of increasing the number of verifications to be made. Moreover, the theory of bimodules incorporates those of left modules and right modules, since a left R-module may be regarded as an $(R,\mathbf{Z})$-bimodule, and a right S-module as a $(\mathbf{Z},S)$-bimodule. Thus the theory of this section is immediately applicable to left or right modules. As in Section 2, we identify an (R,S)-bimodule with an (S^0,R^0)-bimodule, so that the categories $_R\mathfrak{M}_S$, $_{S^0}\mathfrak{M}_{R^0}$ become canonically equivalent. By the same token, it is unnecessary to consider separately (R,S)-*left modules*, meaning abelian groups A that are left R-modules and left S-modules, subject to

$$r(sa) = s(ra).$$

For such an object is just an (R,S^0)-bimodule. It is, in fact, the case that an (R,S)-bimodule may be regarded as a left $(R\times S^0)$-module, but, though this point of view reduces the study of bimodules to that of modules, it is not convenient for us to adopt it.

Now let $A\in {}_R\mathfrak{M}_S, B\in {}_R\mathfrak{M}_T$ and consider the abelian group $\mathrm{Hom}_R(A,B)$. We give $\mathrm{Hom}_R(A,B)$ the structure of an (S,T)-bimodule as follows. Let $\phi\in \mathrm{Hom}_R(A,B)$. Then we define

$$\begin{cases} (s\phi)(a) = \phi(as) \\ (\phi t)(a) = (\phi a)t, \quad a\in A, \quad s\in S, \quad t\in T. \end{cases} \tag{3.2}$$

We check that $\mathrm{Hom}_R(A,B)$ is an S-module. We have

$$s(\phi+\psi)(a) = (\phi+\psi)(as)$$
$$= \phi(as)+\psi(as)$$
$$= (s\phi)(a)+(s\psi)(a)$$
$$= (s\phi+s\psi)(a),$$

so that $\mathbf{M}_1$ is satisfied. Next

$$((s_1+s_2)\phi)(a) = \phi(a(s_1+s_2))$$
$$= \phi(as_1+as_2)$$
$$= \phi(as_1)+\phi(as_2)$$
$$= (s_1\phi)(a)+(s_2\phi)(a)$$
$$= (s_1\phi+s_2\phi)(a),$$

so that $\mathbf{M}_2$ is satisfied. Next

$$((s_1s_2)\phi)(a) = \phi(a(s_1s_2))$$

$$= \phi((as_1)s_2)$$

$$= (s_2\phi)(as_1)$$

$$= (s_1(s_2\phi))(a),$$

so that $\mathbf{M}_3$ is satisfied. $\mathbf{M}_4$ is trivial.

Similarly one checks that $\mathrm{Hom}_R(A, B)$ is a right T-module. To check the compatibility condition we observe that

$$((s\phi)t)(a) = \phi(as)t = (s(\phi t))(a).$$

THEOREM 3.1. *Let* $A \in {}_R\mathfrak{M}_S, B \in {}_R\mathfrak{M}_T$. *Under* (3.2), $\mathrm{Hom}_R(A, -)$ *is a functor* ${}_R\mathfrak{M}_T \to {}_S\mathfrak{M}_T$, *and* $\mathrm{Hom}_R(-, B)$ *is a functor* ${}_R\mathfrak{M}_S^0 \to {}_S\mathfrak{M}_T$.

PROOF. We are content to prove the first assertion. The essence of the claim is that, if $\psi : B_1 \to B_2$ is a morphism of ${}_R\mathfrak{M}_T$, then the function $\phi \mapsto \psi\phi$, $\phi \in \mathrm{Hom}_R(A, B_1)$, is a morphism of ${}_S\mathfrak{M}_T$. Thus we must show that $\psi(s\phi) = s(\psi\phi)$ and $\psi(\phi t) = (\psi\phi)t$.

Now $\psi(s\phi)(a) = \psi(\phi(as)) = (\psi\phi)(as) = (s(\psi\phi))(a)$, and $\psi(\phi t)(a) = \psi(\phi(a)t)$ $= (\psi(\phi(a)))t = ((\psi\phi)(a))t = ((\psi\phi)t)(a)$. This proves the first assertion; notice that, in the last line, we needed the fact that ψ is a T-module map.

There is a parallel construction if one considers $\mathrm{Hom}_R(A, B)$, where $A \in {}_S\mathfrak{M}_R, B \in {}_T\mathfrak{M}_R$. However, it is unnecessary to pursue this in detail, since we may refer back to what we have already done by means of the identifications

$$_S\mathfrak{M}_R = {}_{R^0}\mathfrak{M}_{S^0}, \quad {}_T\mathfrak{M}_R = {}_{R^0}\mathfrak{M}_{T^0}.$$

For then $\mathrm{Hom}_R(A, B) = \mathrm{Hom}_{R^0}(A, B)$ is an (S^0, T^0)-bimodule, that is a (T, S)-module and we have

THEOREM 3.1'. *Let* $A \in {}_S\mathfrak{M}_R, B \in {}_T\mathfrak{M}_R$. *Then* $\mathrm{Hom}_R(A, -)$ *may be made into a functor* ${}_T\mathfrak{M}_R \to {}_T\mathfrak{M}_S$, *and* $\mathrm{Hom}_R(-, B)$ *may be made into a functor* ${}_S\mathfrak{M}_R^0 \to {}_T\mathfrak{M}_S$.

Notice that the formulae corresponding to (3.2) are these: here $A \in {}_S\mathfrak{M}_R, B \in {}_T\mathfrak{M}_R, \phi \in \mathrm{Hom}_R(A, B)$.

$$\begin{cases} (t\phi)(a) = t(\phi a), \\ (\phi s)(a) = \phi(sa), \quad a \in A, \quad s \in S, \quad t \in T. \end{cases} \qquad (3.2')$$

These formulae are more convenient to handle than (3.2); this is due to our convention of writing functions on the *left* of their arguments.

THEOREM 3.2.   *Regard R as an (R, R)-bimodule in the obvious way. Then if* $M \in {}_R\mathfrak{M}_S$, $Hom_R(R, M) \in {}_R\mathfrak{M}_S$, *and there is a natural isomorphism*

$$M \cong \operatorname{Hom}_R(R, M). \tag{3.3}$$

PROOF.   We define $\kappa : \operatorname{Hom}_R(R, M) \to M$ by $\kappa(\phi) = \phi(1)$. It is plain that κ is a bijection and $\kappa(\phi + \psi) = \kappa\phi + \kappa\psi$. We must show that κ is a bimodule map. Now

$$\kappa(r\phi) = (r\phi)(1) = \phi(r) = r(\phi(1)) = r(\kappa(\phi)),$$

and

$$\kappa(\phi s) = (\phi s)(1) = (\phi 1)s = (\kappa(\phi))s.$$

The naturality of κ is obvious.

THEOREM 3.3.   *Let* $A \in {}_R\mathfrak{M}_S$, $B_i \in {}_R\mathfrak{M}_T, i \in I$. *Then there is a natural isomorphism of* (S, T)-*bimodules*

$$\operatorname{Hom}_R(A, \Pi B_i) \cong \Pi \operatorname{Hom}_R(A, B_i). \tag{3.4}$$

PROOF.   Let $p_i : \Pi B_i \to B_i$ be the projection. Plainly $\phi \mapsto \{p_i\phi\}$ defines the required natural isomorphism.

A companion result is

THEOREM 3.4.   *Let* $A_i \in {}_R\mathfrak{M}_S$, $i \in I$, $B \in {}_R\mathfrak{M}_T$. *Then there is a natural isomorphism of* (S, T)-*bimodules*

$$\operatorname{Hom}_R(\oplus A_i, B) \cong \Pi \operatorname{Hom}_R(A_i, B). \tag{3.5}$$

PROOF.   Let $\iota_i : A_i \to \oplus A_i$ be the injection. Plainly $\phi \mapsto \{\phi\iota_i\}$ defines the required natural isomorphism.

THEOREM 3.5
   (*i*) *The functor* $Hom_R(A, -)$ *is left exact.*
   (*ii*) *The functor* $Hom_R(-, B)$ *is left exact.*
   The proof is (naturally) exactly as for abelian groups (Theorem 4.2 of Chapter 2).

4. THE FUNCTOR $\otimes$

In this section we construct a left adjoint to the Hom functor; this will be very important in the sequel. Our definition generalizes that of Chapter 2.

DEFINITION 4.1.   Let $A \in {}_R\mathfrak{M}_S$, $B \in {}_S\mathfrak{M}_T$. We define $A \otimes_S B \in {}_R\mathfrak{M}_T$, the *tensor product* of A and B over S, as follows. We form the tensor product $A \otimes B$ of the abelian groups A, B and adjoin the relations

$$as \otimes b = a \otimes sb, \quad a \in A, \quad b \in B, \quad s \in S. \tag{4.1}$$

The resulting abelian group $A \otimes_S B$ is then given the structure of an (R, T)-bimodule by the rules

$$r(a \otimes b) = ra \otimes b, \quad (a \otimes b)t = a \otimes bt, \quad r \in R, \quad t \in T. \tag{4.2}$$

Recall that $A \otimes B$ is the abelian group generated by symbols $a \otimes b$, subject to the relations

$$(a_1 + a_2) \otimes b = a_1 \otimes b + a_2 \otimes b, \quad a \otimes (b_1 + b_2) = a \otimes b_1 + a \otimes b_2. \tag{4.3}$$

Thus, to justify this definition, we must show that the rules (4.2) respect the relations (4.1) and (4.3). It is clear that they respect (4.3). As to (4.1), we have

$$r(as \otimes b) = r(as) \otimes b = (ra)s \otimes b = ra \otimes sb = r(a \otimes sb),$$

and

$$(as \otimes b)t = as \otimes bt = a \otimes s(bt) = a \otimes (sb)t = (a \otimes sb)t.$$

Notice that, in this verification, the compatibility condition (3.1) is crucial.

Of course, we need not write $A \otimes_{\mathbf{Z}} B$, if A is a right $\mathbf{Z}$-module (abelian group) and B a left $\mathbf{Z}$-module (abelian group), since in this case the relation (4.1) is automatic. Thus we continue to write $A \otimes B$ if a tensor product over $\mathbf{Z}$ is in question.

We need not repeat the elementary properties of tensor products already proved in Chapter 2. Further elementary properties are listed below.

PROPOSITION 4.1
(i) $A_1 \otimes_R A_2 \cong A_2 \otimes_{R^o} A_1$.
(ii) If $A \in {}_R\mathfrak{M}_S, B \in {}_S\mathfrak{M}_T, C \in {}_T\mathfrak{M}_U$, then $(A \otimes_S B) \otimes_T C \cong A \otimes_S(B \otimes_T C)$.
We leave the proof as an exercise.

PROPOSITION 4.2.   *If* $A \in {}_R\mathfrak{M}_S$ *and* R *is regarded as an* (R, R)-*bimodule, then* $R \otimes_R A \cong A$ *under*

$$r \otimes a \mapsto ra. \tag{4.4}$$

PROOF.   It is plain that the map $\theta : R \otimes_R A \to A$, given by (4.4), is a well-defined map of (R, S)-bimodules. Then θ has an inverse given by $a \mapsto 1 \otimes a$, and so is an isomorphism.

It is very easy to turn $A \otimes_S -, - \otimes_S B$ into functors, for $A \in {}_R\mathfrak{M}_S$, $B \in {}_S\mathfrak{M}_T$. Indeed, if $\alpha : A \to A'$ in ${}_R\mathfrak{M}_S$, $\beta : B \to B'$ in ${}_S\mathfrak{M}_T$, there is a well-defined function $\alpha \otimes \beta : A \otimes_S B \to A' \otimes_S B'$, given by

$$(\alpha \otimes \beta) \sum_i a_i \otimes b_i = \sum_i \alpha a_i \otimes \beta b_i. \tag{4.5}$$

The reader should check (it is quite routine) that (4.5) respects the defining relations for the tensor product over S; moreover, (4.5) is plainly a homomorphism of (R, T)-bimodules, and so we define the functor $A \otimes_S -$ by

$$A \otimes_S \beta = 1_A \otimes \beta, \tag{4.6}$$

and the functor $- \otimes_S B$ by

$$\alpha \otimes_S B = \alpha \otimes 1_B. \tag{4.7}$$

Again, it is plain that we do thereby obtain functors; indeed, we obtain additive functors. Notice that $A \otimes_S - : {}_S\mathfrak{M}_T \to {}_R\mathfrak{M}_T$, and $- \otimes_S B : {}_R\mathfrak{M}_S \to {}_R\mathfrak{M}_T$.

We now come to a crucial theorem in the theory of modules.

THEOREM 4.3. *Let* B *belongs to* ${}_S\mathfrak{M}_T$. *Then* $- \otimes_S B : {}_R\mathfrak{M}_S \to {}_R\mathfrak{M}_T$ *is left adjoint to* $\text{Hom}_T(B, -) : {}_R\mathfrak{M}_T \to {}_R\mathfrak{M}_S$.

PROOF. We define, for $A \in {}_R\mathfrak{M}_S, C \in {}_R\mathfrak{M}_T$,

$$\eta : \text{Hom}_{(R,T)}(A \otimes_S B, C) \cong \text{Hom}_{(R,S)}(A, \text{Hom}_T(B, C))$$

by the rule

$$(\eta(\theta)(a))(b) = \theta(a \otimes b), \quad \theta \in \text{Hom}_{(R,T)}(A \otimes_S B, C). \tag{4.8}$$

Now $\eta(\theta)(a) \in \text{Hom}_T(B, C)$, since

$$(\eta(\theta)(a))(bt) = \theta(a \otimes bt) = \theta((a \otimes b)t) = (\theta(a \otimes b))t = ((\eta(\theta)(a))(b))t.$$

Next $\eta(\theta) \in \text{Hom}_{(R,S)}(A, \text{Hom}_T(B, C))$, since

$$(\eta(\theta)(ra))(b) = \theta(ra \otimes b)$$

$$= \theta(r(a \otimes b))$$

$$= r\theta(a \otimes b)$$

$$= r(\eta(\theta)(a)(b))$$

$$= (r(\eta(\theta)(a)))(b), \quad \text{by (3.2}')$$

so that

$$\eta(\theta)(ra) = r(\eta(\theta)(a));$$

and

$$(\eta(\theta)(as))(b) = \theta(as \otimes b)$$
$$= \theta(a \otimes sb)$$
$$= (\eta(\theta)(a))(sb)$$
$$= ((\eta(\theta)(a))s)(b) \quad \text{by } (3.2')$$

so that

$$\eta(\theta)(as) = (\eta(\theta)(a))s.$$

Conversely, given $\bar{\theta} \in \text{Hom}_{(R,S)}(A, \text{Hom}_T(B, C))$, we may define $\bar{\eta}(\bar{\theta}) \in \text{Hom}_{(R,T)}(A \otimes_S B, C)$ by the rule

$$\bar{\eta}(\bar{\theta})(a \otimes b) = \bar{\theta}(a)(b). \tag{4.9}$$

We leave it to the reader to verify that $\bar{\eta}(\bar{\theta}) \in \text{Hom}_{(R,T)}(A \otimes_S B, C)$ and that $\bar{\eta}$ is inverse to η. Likewise, the naturality of η is so obvious that we do not need to write out the argument.

Notes

1. We have not troubled to show that η and $\bar{\eta}$ preserve abelian group structure, since this is an immediate comsequence of (4.3). The only subtlety lies in the preservation of the various ring actions.

2. Of course, there is an alternative form of this theorem using (3.2) instead of (3.2'). It is immediately deducible from the statement of the theorem, using the standard trick of passing to opposite rings.

We draw some immediate consequences.

COROLLARY 4.4. *The functors* $A \otimes_S -, - \otimes_S B$ *are right exact.*

PROOF. We showed in Section 7 of Chapter 3 that *any* functors with right adjoints were right exact.

COROLLARY 4.5. *The functors* $A \otimes_S -, - \otimes_S B$ *preserve direct sums.*

PROOF. We showed in Section 7 of Chapter 3 that *any* functors with right adjoints preserve coproducts.

THEOREM 4.6. *Let* F *be a right* R-*module, isomorphic to a direct sum of copies of* R; *thus*

$$F \cong \bigoplus_i R.$$

Then, if $A \in {}_R\mathfrak{M}_S, F \otimes_R A \cong \oplus_i A$, *as right* S-*modules.*

PROOF.   Apply Corollary 4.5 and Proposition 4.2.

Of course, this theorem could also be stated for left modules.

5. PROJECTIVE MODULES

In the next two sections we generalize concepts defined in Chapter 2, using the categorical notions of Chapter 3. We first define the important concept of a *free* left R-module.

DEFINITION 5.1.   Let S be a set. The *free left* R-*module on the set* S is a pair consisting of a left R-module F and a function $j : S \to F$, enjoying the following universal property. For any left R-module A and function $f : S \to A$, there exists a unique R-module homomorphism $\phi : F \to A$ with $\phi j = f$.

PROPOSITION 5.1.   *The free module* F *is generated by* j(S).

PROOF.   Let F' be the submodule of F generated by $j(S)$. If $F' \neq F$, then $\pi \neq 0 : F \to F/F'$. But $\pi j = 0 j$, since $j(S) \subseteq F'$. This contradicts the uniqueness of ϕ in the definition of a free module.

We next point out that the free module on S (if it exists) is unique.

PROPOSITION 5.2.   *The free left* R-*module on* S *is unique up to canonical isomorphism. Precisely, if* (F,j) *and* (F′,j′) *are both free modules on* S, *there exists a unique isomorphism* $\theta : F \cong F'$, *with* $\theta j = j'$.

We leave the proof to the reader. It now remains to establish the existence of free modules. Let S be a set, let $\bar{F} = \oplus_{s \in S} R$, regarded as a left R-module, and let $\bar{j} : S \to \bar{F}$ be given by

$$\bar{j}(s) = 1_s, \quad s \in S, \tag{5.1}$$

where 1_s stands for the element of $\bar{F}$ whose tth component, $t \in S$, is the Kronecker symbol δ_{st}. (Thus the sth component of 1_s is 1, the other components are zero.)

PROPOSITION 5.3. $(\bar{F}, \bar{j})$ *is the free left* R-*module on* S.

PROOF. Let $f: S \to A$ be a function from S to the module A. We define $\phi: \bar{F} \to A$ by the rule

$$\phi(1_s) = f(s), \quad s \in S.$$

This plainly yields a homomorphism $\phi: \bar{F} \to A$ with $\phi \bar{j} = f$. Morever, $\bar{j}(S)$ generates $\bar{F}$, so that, by Corollary 2.10, ϕ is uniquely determined by the equation $\phi \bar{j} = f$.

We note that $\bar{j}$ is injective; so therefore is the function j in any representation of the free module (F, j) on the set S. Thus we always suppose S to be embedded in F by means of j, so that S is a set of generators of F. We call S an *R-basis* or, simply, *basis* of the free module F.

PROPOSITION 5.4. *Let* S *be a set of generators of the* R-*module* B. *Then* B *is the free module on the basis* S *if and only if the set* S *is linearly independent over* R, *that is, if and only if the relation*

$$\sum r_i s_i = 0 \quad \text{(finite sum)}$$

implies that each $r_i = 0$.

PROOF. Let $\psi: \bar{F} \to B$ be the homomorphism given by

$$\psi(1_s) = s.$$

We have to show that ψ is an isomorphism if and only if S is linearly independent over R. We note that (i) ψ is surjective since S generates B and (ii) the set $\bar{S}$ of elements 1_s in $\bar{F}$ is linearly independent over R. It is now obvious that ψ is injective if and only if S is linearly independent over R.

Thus we say that the R-module B is *free* if and only if it has an R-basis, S. If so, B is then *free on* S.

THEOREM 5.5. *Every* R-*module is the homomorphic image of a free* R-*module.*

PROOF. Let A be an R-module and let S be a generating set of A (for example, we could take $S = A$ or $S = A - (0)$). Let F be the free module on S and let $\phi: F \to A$ be defined by $\phi | S = i$, the inclusion of S in A. Obviously ϕ is surjective, since S generates A.

We come now to the principal definition of this section.

DEFINITION 5.2. A left R-module P is said to be *projective* if, given any surjection $\epsilon: A \to B$ in $_R\mathfrak{M}$ and homomorphism $\phi: P \to B$, there exists $\psi: P \to A$ with $\epsilon \psi = \phi$.

PROPOSITION 5.6. *A direct sum $\oplus_i P_i$ is projective if and only if each P_i is projective.*

PROOF. Suppose that $\oplus_i P_i$ is projective and let $\epsilon : A \twoheadrightarrow B$, $\phi_{i_0} : P_{i_0} \to B$ be given. Set $\phi_i = 0 : P_i \to B, i \neq i_0$, and let $\phi = \langle \phi_i \rangle : \oplus_i P_i \to B$. Since $\oplus_i P_i$ is projective, there exists $\psi = \langle \psi_i \rangle : \oplus_i P_i \to A$, with $\epsilon \psi = \phi$. Plainly $\epsilon \psi_{i_0} = \phi_{i_0}$.

Conversely, suppose each P_i projective, and let $\epsilon : A \twoheadrightarrow B, \phi = \langle \phi_i \rangle : \oplus_i P_i \to B$ be given. Then, for each i, there exists $\psi_i = P_i \to A$ with $\epsilon \psi_i = \phi_i$. Thus $\epsilon \psi = \phi$ with $\psi = \langle \psi_i \rangle$.

THEOREM 5.7. *A free module is projective.*

PROOF. Give $\epsilon : A \twoheadrightarrow B$ and $\phi : F \to B$, where F is free on S, choose, for each $s \in S$, an element $a_s \in A$ with $\epsilon(a_s) = \phi s$, and let $f : S \to A$ be given by $fs = a_s$. Let $\psi : F \to A$ be determined by $\psi|S = f$. Then $\epsilon \psi|S = \phi|S$, so that, by uniqueness, $\epsilon \psi = \phi$.

On the other hand, there are, in general, projective modules that are not free. For example, one may readily show that

$$\mathbf{Z}_6 \cong \mathbf{Z}_2 \oplus \mathbf{Z}_3$$

as $\mathbf{Z}_6$-modules. Since $\mathbf{Z}_6$ is, of course, a free $\mathbf{Z}_6$-module, it follows from Proposition 5.6 and Theorem 5.7 that $\mathbf{Z}_2$ and $\mathbf{Z}_3$ are projective $\mathbf{Z}_6$-modules, but they are obviously not free.

PROPOSITION 5.8. *The R-module P is projective if and only if the functor $Hom_R(P, -) : {}_R\mathfrak{M} \to \mathfrak{Ab}$ is exact.*

PROOF. This is immediate from Theorem 3.5(i) and Definition 5.2.

We now characterize projective modules in two useful ways.

THEOREM 5.9. *The following statements about the R-module P are equivalent:*
  (*i*) *P is projective.*
  (*ii*) *Every short exact sequence $A' \rightarrowtail A \twoheadrightarrow P$ splits.*
  (*iii*) *P is a direct summand in a free module.*

PROOF

(i)$\Rightarrow$(ii) Given $A' \overset{\mu}{\rightarrowtail} A \overset{\epsilon}{\twoheadrightarrow} P$, there exists, since P is projective, $\psi : P \to A$ with $\epsilon \psi = 1$.

(ii)⇒(iii) By Theorem 5.5 we may find $F \xrightarrow{\epsilon} P$ with F free. Let K be the kernel of F. Then $K \rightarrowtail F \xrightarrow{\epsilon} P$ is short exact and so it splits. Thus $F \cong P \oplus K$.

(iii)⇒(i) Theorem 5.7 and Proposition 5.6. (Compare Theorem 8.14 in Chapter 3.)

We now look at the relation of projective modules to tensor products. We say that the R-module B is *flat* if the functor $- \otimes_R B$ is exact (see Corollary 4.4). This is equivalent to demanding that $- \otimes_R B$ preserve monomorphisms.

THEOREM 5.10. *Projective modules are flat.*

PROOF. We first prove that free modules are flat. Let F be the free module on S; thus $F = \oplus_{s \in S} R$. Then $A \otimes_R F \cong \oplus_{s \in S} A$ (Theorem 4.6), so that, if μ is the embedding of A' in $A, \mu \otimes_R F$ is just the embedding of $\oplus_s A'$ in $\oplus_s A$.

Now let P be projective. Then $P \oplus Q = F$, free, for some Q, by Theorem 5.9. Let $\iota : P \to F, \pi : F \to P$, be the injection and projection respectively. Consider the diagram

$$
\begin{array}{ccc}
A' \otimes_R P & \xrightarrow{\mu_*} & A \otimes_R P \\
\downarrow{\iota_*} & & \downarrow{\iota_*} \\
A' \otimes_R F & \xrightarrow{\mu_*} & A \otimes_R F \\
\downarrow{\pi_*} & & \downarrow{\pi_*} \\
A' \otimes_R P & \xrightarrow{\mu_*} & A \otimes_R P
\end{array}
$$

where we have written μ_*, ι_*, π_* to avoid tiresome functorial symbols. We want to show that $\mu_* : A' \otimes_R P \to A \otimes_R P$ is injective. We know that $\mu_* : A' \otimes_R F \to A \otimes_R F$ is injective and that $\pi_* \iota_* = 1$. Let $x \in A' \otimes_R P$, $\mu_* x = 0$. Then $0 = \iota_* \mu_* x = \mu_* \iota_* x$, so that $\iota_* x = 0$. Thus $x = \pi_* \iota_* x = 0$.

Again the converse is false. For one may easily show that a module is flat if all its finitely generated submodules are flat. Now a torsion-free abelian group has all its finitely generated submodules free, hence flat. However, since a subgroup of a free abelian group is free, it follows from Theorem 5.9 that every projective abelian group is free. Thus we disprove the converse by observing that there are torsion-free abelian groups that are not free, for example, the group **Q** of rationals.

We now proceed to an important characterization of projective modules, which does not lend itself so readily to categorical formulation. If A is a left R-module, then, as shown in Section 3, $A^* = \mathrm{Hom}_R(A, R)$ acquires the structure of a right R-module by the rule

$$(\phi r)(a) = \phi(a)r. \tag{5.1}$$

We call A^* the *dual* module of A. In the next theorem, however, we do not need the module structure of A^*.

THEOREM 5.11. *The left R-module* P *is projective if and only if there are subsets* $\{x_i\}, i \in I$, *of* P, *and* $\{f_i\}, i \in I$, *of* P^*, *such that, for each* $x \in P, f_i(x) = 0$ *for almost all* i, *and*

$$x = \sum f_i(x) x_i.$$

PROOF. Let P be projective and $F = P \oplus Q$, with F free. Let $\{z_i\}$ be an R-basis for F, with $z_i = x_i + y_i, x_i \in P, y_i \in Q$. Let $g_i : F \to R$ be given by $g_i(z_j) = \delta_{ij}$, and let $f_i = g_i | P$. For each $x \in P, x = \sum r_i z_i = \sum r_i x_i$, where $r_i = 0$ for almost all i. Plainly if $r_i = 0, f_i(x) = 0$, since $f_i(x) = r_i$. The same argument shows that $x = \sum f_i(x) x_i$.

Conversely, suppose that we are given the subsets $\{x_i\}, \{f_i\}$, and let F be the free left R-module on a set $\{z_i\}$, in one-one correspondence with $\{x_i\}$. We thus have $\phi : F \to P$ given by $\phi(z_i) = x_i$. However, we claim that the function $\psi : P \to F$, given by $\psi(x) = \sum f_i(x) z_i$ is a homomorphism. For

$$\psi(x+y) = \sum f_i(x+y) z_i = \sum (f_i(x) + f_i(y)) z_i$$

$$= \sum f_i(x) z_i + \sum f_i(y) z_i = \psi(x) + \psi(y),$$

and

$$\psi(rx) = \sum f_i(rx) z_i = \sum (rf_i(x)) z_i = r \sum f_i(x) z_i = r\psi(x).$$

Moreover, $\phi\psi(x) = \sum f_i(x) x_i = x, \phi\psi = 1$, so that P is a direct summand in F, hence projective.

We call the sets $\{x_i\}, \{f_i\}$ an *R-projective basis* for P. Notice that P is generated by $\{x_i\}$.

COROLLARY 5.12. *The left R-module* P *is finitely-generated projective if and only if there are finite subsets* $\{x_i\}$ *of* P, $\{f_i\}$ *of* P^* *such that* $x = \sum f_i(x) x_i$ *for all* $x \in P$.

PROOF. If the condition holds, then P is projective by Theorem 5.11 and finitely generated, since $\{x_i\}$ is a generating set. Conversely, if P is finitely generated projective, we apply the proof of Theorem 5.11, refined by the observation that if P is finitely generated, we may take F to have a finite R-basis.

We note that if F is a free finitely generated left R-module, then F^* is a free finitely generated right R-module. Thus if P is a finitely generated projective left R-module, P^* is a finitely generated projective right R-module. Moreover, we easily prove (identifying P^{**} with P in the obvious way).

THEOREM 5.13. *If* $\{x_i\}, \{f_i\}$ *is a finite* R-*projective basis for* P, *then* $\{f_i\}, \{x_i\}$ *is a finite* R-*projective basis for* P^*.

PROOF. We have only to prove that, for any $f \in P^*$,

$$f = \sum f_i f(x_i). \tag{5.2}$$

Now let $g = \sum f_i f(x_i)$. Then

$$g(x) = \sum (f_i f(x_i))(x) = \sum f_i(x) f(x_i), \quad \text{by (5.1)}$$
$$= f\left(\sum f_i(x) x_i\right) = f(x).$$

Our next result will be crucial in Chapter 6. Let $A, B \in {}_R\mathfrak{M}$, and define

$$\theta_{AB} : A^* \otimes_R B \to \operatorname{Hom}_R(A, B)$$

by

$$\theta(f \otimes b)(a) = f(a)b. \tag{5.3}$$

Then θ is plainly well-defined; we are content to check explicitly that

$$\theta(fr \otimes b)(a) = ((fr)(a))b = ((fa)r)b = (fa)rb = (f \otimes rb)(a).$$

Moreover $\theta(f \otimes b)$ is a homomorphism of R-modules, since

$$\theta(f \otimes b)(ra) = f(ra)b = rf(a)b = r\theta(f \otimes b)(a).$$

Indeed it is plain that θ is a natural transformation of functors ${}_R\mathfrak{M}^0 \times {}_R\mathfrak{M} \to \mathfrak{Ab}$; furthermore, it is easy to show that we may even regard θ as a natural transformation of functors ${}_R\mathfrak{M}_S^0 \times {}_R\mathfrak{M}_T \to {}_S\mathfrak{M}_T$. However, we do not insist for the time being on this generality. We prove[†]

THEOREM 5.14. *The natural transformation* $\theta_{A-} : A^* \otimes_R - \to \operatorname{Hom}_R(A, -)$ *is an isomorphism if and only if* A *is finitely generated projective.*

PROOF. Suppose that θ_{A-} is an isomorphism. Then, in particular, $\theta_{AA} : A^* \otimes_R A \cong \operatorname{Hom}_R(A, A)$. Suppose now that

$$\theta\left(\sum_{i=1}^n f_i \otimes x_i\right) = 1, \quad f_i \in A^*, \quad x_i \in A.$$

[†]This theorem is of great importance in algebraic topology, in connection with the universal coefficient theorems in cohomology.

Then $\sum_{i=1}^{n} f_i(x)x_i = \theta(\sum_{i=1}^{n} f_i \otimes x_i)(x) = x$, so that $\{x_i\}, \{f_i\}$ is a finite projective R-basis for A.

Conversely, we observe that θ_{A-} is obviously an isomorphism if A is free on a finite R-basis S. For then $A = \oplus_s R$, whence $A^* = \oplus_s R$ as a right R-module, $A^* \otimes_R B \cong \oplus_s B$, $\text{Hom}_R(A, B) \cong \oplus_s B$, and θ is plainly the isomorphism here referred to between $A^* \otimes_R B$ and $\text{Hom}_R(A, B)$. It is now only necessary to observe that θ respects finite direct sums in the first variable to infer that θ_{A-} is an isomorphism if A is finitely generated projective. (Of course θ also respects finite direct sums in the second variable.)

From this theorem we are able to infer a partial converse to Theorem 5.10. We say that a module A is *finitely presented* if there is an exact sequence

$$F_1 \to F_0 \to A \to 0$$

with both F_0 and F_1 free finitely generated. Then we may prove

THEOREM 5.15.   *A finitely presented flat module is projective.*

PROOF.   Let $F_1 \to F_0 \to A \to 0$ be a presentation of A with F_0, F_1 free finitely generated. We obtain the commutative diagram

$$
\begin{array}{ccccc}
0 \longrightarrow & A^* \otimes_R A & \longrightarrow & F_0^* \otimes_R A & \longrightarrow & F_1^* \otimes_R A \\
& \downarrow{\theta_{AA}} & & \downarrow{\theta_{F_0 A}} & & \downarrow{\theta_{F_1 A}} \\
0 \longrightarrow & \text{Hom}_R(A, A) & \longrightarrow & \text{Hom}_R(F_0, A) & \longrightarrow & \text{Hom}_R(F_1, A)
\end{array}
\quad (5.4)
$$

Now since $\text{Hom}_R(-, R)$ is left exact, $0 \to A^* \to F_0^* \to F_1^*$ is exact. Since A is flat, the top row of (5.4) is exact. Since $\text{Hom}_R(-, A)$ is left exact, the bottom row of (5.4) is exact. By Theorem 5.14, $\theta_{F_0 A}$ and $\theta_{F_1 A}$ are isomorphisms; so therefore, by standard reasoning (see, for example, the proof of Proposition 7.3 in Chapter 1) is θ_{AA}. However, we saw in the proof of Theorem 5.14 that the fact that θ_{AA} is an isomorphism is sufficient to allow us to deduce that A is (finitely generated) projective.

We close this section with a useful theorem on the transfer of projectives from one category of modules to another. Although this theorem admits several different proofs, we prefer to give a categorical proof that illustrates the general principle governing the result.

Let $\theta : R \to S$ be a ring homomorphism. Then every S-module may be regarded as an R-module via θ, as explained in Example 6 of Section 2. We obtain in this way a functor $U : {}_S\mathfrak{M} \to {}_R\mathfrak{M}$.

THEOREM 5.16. *Let* S *be projective as* R-*module. Then every projective* S-*module is also a projective* R-*module.*

PROOF. The functor U is left adjoint to the functor $F: {}_R\mathfrak{M} \to {}_S\mathfrak{M}$, given by

$$FA = \operatorname{Hom}_R(S,A).$$

Since S is a projective R-module, F is an exact functor (see Proposition 5.8). In particular, F preserves epics.

Now let P be a projective S-module and consider the diagram in ${}_R\mathfrak{M}$,

$$
\begin{array}{c}
UP \\
\downarrow \phi \\
A \xrightarrow{\ \epsilon\ } B
\end{array}
$$

Transferring by the adjunction η to ${}_S\mathfrak{M}$, we get

$$
\begin{array}{c}
P \\
\downarrow \eta(\phi) \\
FA \xrightarrow{\ F\epsilon\ } FB
\end{array}
$$

Since P is projective and $F\epsilon$ is epic, we find $\psi': P \to FA$ with $(F\epsilon)\psi' = \eta(\phi)$. Returning to ${}_S\mathfrak{M}$ by the adjunction, we have $\epsilon\psi = \phi$, where $\eta(\psi) = \psi'$. Thus P is projective. This is a special case of the argument of Theorem 8.12 in Chapter 3.

COROLLARY 5.17. *A projective* **Z**[G]-*module is a free abelian group.*

PROOF. As abelian group, **Z**[G] is free on G as basis. Thus a projective **Z**[G]-module is a projective abelian group. But (since subgroups of free abelian groups are free abelian) projective abelian groups are free.

6. INJECTIVE MODULES

In this section we develop a theory that is, in a precise sense, dual to that of projective modules.

DEFINITION 6.1. A left R-module I is said to be *injective* if, given any injection $\mu: B \to A$ in ${}_R\mathfrak{M}$ and homomorphism $\phi: B \to I$, there exists $\psi: A \to I$ with $\psi\mu = \phi$.

Thus (as the diagram instantly reveals) an object of $_R\mathfrak{M}$ is injective if and only if it is a projective object of $_R\mathfrak{M}^0$; we can therefore expect to find analogs of many of the facts about projective modules.

PROPOSITION 6.2. *A direct product* $\prod_i I_i$ *is injective if and only if each* I_i *is injective.*

This is dual to Proposition 5.6, so we suppress the proof.

PROPOSITION 6.3. *The R-module* I *is injective if and only if the functor* $Hom_R(-,I): _R\mathfrak{M}^0 \to \mathfrak{Ab}$ *is exact.*

This is dual to Proposition 5.8.

Now the search for projective modules was greatly simplified by the presence of free modules. We will obtain a heuristic "dual" to the concept of a free module, which will enable us to find plenty of injective modules. But another procedure is to use the data of Theorem 5.16. We recall that we have already identified the injective abelian groups—they are simply the *divisible* abelian groups. We suppose that we are given, as in Theorem 5.16, a ring homomorphism $\theta: R \to S$.

THEOREM 6.4. *Let* I *be an injective R-module. Then* $Hom_R(S,I)$ *is an injective S-module.*

PROOF. In fact, we apply Theorem 8.12* of Chapter 3, but we give the details of the argument. We consider the functors $U: _S\mathfrak{M} \to _R\mathfrak{M}$, $F: _R\mathfrak{M} \to _S\mathfrak{M}$ as in Theorem 5.16. Then $FI = Hom_R(S,I)$, and U is an exact functor. Now suppose that we are given, in $_S\mathfrak{M}$, with I injective,

$$FI$$
$$\uparrow \phi$$
$$A \longleftarrow\!\!\!< B$$
$$\mu$$

Transferring by the adjugant η^{-1} to $_R\mathfrak{M}$, we get

$$I$$
$$\uparrow \eta^{-1}(\phi)$$
$$UA \longleftarrow\!\!\!< UB$$
$$U\mu$$

Thus there exists $\psi' : UA \to I$ with $\psi'(U\mu) = \eta^{-1}(\phi)$. Returning to $_S\mathfrak{M}$ via η, we find $\psi\mu = \phi$, where $\psi = \eta(\psi')$.

COROLLARY 6.5. *Let* D *be a divisible abelian group. Then* Hom(R, D) *is an injective* R-*module.*

Now Theorems 5.5 and 5.7 imply, of course, that every module is the homomorphic image of a projective module. Corollary 6.5 enables us to dualize this result to show that every module may be embedded in an injective module: this result is due originally to Baer, but our proof is due essentially to Eckmann-Schopf.

THEOREM 6.6. *Every* R-*module may be embedded in an injective* R-*module.*

PROOF Let A be an R-module, B an abelian group, $U : {}_R\mathfrak{M} \to \mathfrak{A}\mathfrak{b}$ the underlying functor with right adjoint $F = \text{Hom}(R, -)$. Then $\text{Hom}(UA, B) \cong \text{Hom}_R(A, \text{Hom}(R, B))$. If we take $B = UA$, it is easy to see that corresponding to 1_{UA} we obtain

$$\iota : A \to \text{Hom}(R, UA),$$

given by $a \mapsto \phi_a$, where $\phi_a(r) = ra$. Thus ι is monic. Now embed A in the divisible abelian group D (Theorems 7.10 of Chapter 2); we get

$$A \overset{\iota}{\mapsto} \text{Hom}(R, UA) \subseteq \text{Hom}(R, D),$$

which, by Corollary 6.5, is an embedding of A in an injective R-module.

We now approach the question of injective modules from a somewhat different standpoint. Our objective is to get an analog of Theorem 5.9.

Let F be a free right R-module, and let $\mathbf{Q}_1$ be the abelian group

$$\mathbf{Q}_1 = \mathbf{Q}/\mathbf{Z}.$$

We call the left R-module $\text{Hom}(F, \mathbf{Q}_1)$ a *cofree* module. Since $F = \oplus R$ (over some indexing set) and $\mathbf{Q}_1$ is divisible, we immediately have, from Proposition 6.2 and Corollary 6.5,

PROPOSITION 6.7. *A cofree module is injective.*

Now, given any module A, let $A^{\ddagger} = \text{Hom}(A, \mathbf{Q}_1)$. There is then a map $\iota : A \to A^{\ddagger\ddagger}$, given by

$$\iota(a)(\phi) = \phi(a), \quad a \in A, \quad \phi \in A^{\ddagger}.$$

PROPOSITION 6.8. ι *is a module monomorphism.*

PROOF. Suppose for definiteness that A is a left R-module. Then

$$\iota(ra)(\phi) = \phi(ra) = (\phi r)(a) = \iota(a)(\phi r) = (r(\iota(a)))(\phi).$$

Thus $\iota(ra) = r(\iota(a))$, so that ι is a module homomorphism. To show that ι is injective, it is plainly sufficient to show that, if $a \neq 0$, there exists $\phi \in A^{\ddagger}$ with $\phi(a) \neq 0$. Here we need merely consider A as an abelian group. If a is of infinite order, define $\phi(a)$ to be any nonzero element of $\mathbf{Q}_1$; if a is of order n, define $\phi(a) = 1/n \bmod \mathbf{Z}$. Thus ϕ is defined on the cyclic subgroup of A generated by a, and $\phi(a) \neq 0$. Since $\mathbf{Q}_1$ is divisible, hence injective, ϕ may be extended to the whole of A.

We may now improve Theorem 6.6.

PROPOSITION 6.9. *Every module may be embedded in a cofree module.*

PROOF. Let A be a left R-module. Present $A^{\ddagger}$ by $F \xrightarrow{\epsilon} A^{\ddagger}$, with F free. Since $\mathrm{Hom}(-, \mathbf{Q}_1)$ is exact, we have $A^{\ddagger\ddagger} \xrightarrow{\epsilon^{\ddagger}} F^{\ddagger}$, where $\epsilon^{\ddagger} = \mathrm{Hom}(\epsilon, \mathbf{Q}_1)$. Thus

$$A \xrightarrow{\iota} A^{\ddagger\ddagger} \xrightarrow{\epsilon^{\ddagger}} F^{\ddagger}$$

is an embedding of A in the cofree module $F^{\ddagger}$.

We are ready to "dualize" Theorem 5.9.

THEOREM 6.10. *The following statements about the R-module I are equivalent:*

(i) I *is injective.*

(ii) Every short exact sequence $I \rightarrowtail A \twoheadrightarrow A''$ *splits.*

(iii) I *is a direct factor in a cofree module.*

Since every step in the proof of Theorem 5.9 can be dualized, we leave the details to the reader.

Of course, it follows from Propositions 6.2 and 6.7 that, if P is projective, $P^{\ddagger}$ is injective; we improve this in the next theorem.

THEOREM 6.11. *The module* A *is flat if and only if* $A^{\ddagger}$ *is injective.*

PROOF. We require a lemma, which will be strengthened in the Exercises (Exercise 6.7).

LEMMA 6.12. *A morphism* $\mu : B \to B'$ *is monic if and only if* $\mu^{\ddagger} : B'^{\ddagger} \to B^{\ddagger}$ *is epic.*

PROOF OF LEMMA. Since $\mathrm{Hom}(-, \mathbf{Q}_1)$ is exact, we know that $\mu^{\ddagger}$ is epic if μ is monic, and that $\mu^{\ddagger\ddagger}$ is monic if $\mu^{\ddagger}$ is epic. Thus we complete the proof of the lemma by deducing from the commutative diagram

$$
\begin{array}{ccc}
B & \xrightarrow{\mu} & B' \\
\downarrow{\scriptstyle \iota} & & \downarrow{\scriptstyle \iota} \\
B^{\ddagger\ddagger} & \xrightarrow[\mu^{\ddagger\ddagger}]{} & B'^{\ddagger\ddagger}
\end{array}
$$

that μ is monic if $\mu^{\ddagger\ddagger}$ is monic.

We return now to the proof of the theorem. Let $\mu: B \rightarrowtail B'$ be an arbitrary monic (of right R-modules). Then A is flat if and only if $\mu \otimes 1: B \otimes_R A \to B' \otimes_R A$ is monic for all μ. By the lemma, we infer that A is flat if and only if $(\mu \otimes 1)^{\ddagger}: \operatorname{Hom}(B' \otimes_R A, \mathbf{Q}_1) \to \operatorname{Hom}(B \otimes_R A, \mathbf{Q}_1)$ is epic for all μ. By our fundamental adjointness theorem (Theorem 4.3), this means that A is flat if and only if $\mu^*: \operatorname{Hom}_R(B', A^{\ddagger}) \to \operatorname{Hom}_R(B, A^{\ddagger})$ is epic for all μ. But the latter is precisely the definition that $A^{\ddagger}$ is injective.

We close with an analog of Theorem 5.16; again we suppose that we have a ring homomorphism $\theta: R \to S$.

THEOREM 6.13. *Let* S *be flat as an* R-*module. Then an injective* S-*module is an injective* R-*module.*

PROOF. The functor $U: {}_S\mathfrak{M} \to {}_R\mathfrak{M}$ has a left adjoint $F: {}_R\mathfrak{M} \to {}_S\mathfrak{M}$, given by

$$FA = S \otimes_R A.$$

Since S is flat as an R-module, F is exact. We now complete the argument just as for Theorem 5.16. (See Theorem 8.12* of Chapter 3.)

COROLLARY 6.14. *An injective* $\mathbf{Z}[G]$-*module is a divisible abelian group.* The proof is left to the reader.

EXERCISES

1.1. Let R be a nonunitary ring (that is, we do not demand $\mathbf{R}_4$). In the set $R \times \mathbf{Z}$ introduce addition and multiplication by the rules

$$(r_1, n_1) + (r_2, n_2) = (r_1 + r_2, n_1 + n_2)$$

$$(r_1, n_1)(r_2, n_2) = (r_1 r_2 + n_2 r_1 + n_1 r_2, n_1 n_2).$$

Let us write $\bar{R}$ for the triple $(R \times \mathbf{Z}; +, \times)$ thus defined. Show that $\bar{R}$ is a unitary ring. Let $\mathfrak{R}, \mathfrak{R}_1$ be the categories of nonunitary and unitary rings respectively. Show that $R \mapsto \bar{R}$ yields a functor $F: \mathfrak{R} \to \mathfrak{R}_1$, and that F is left adjoint to the underlying functor $U: \mathfrak{R}_1 \to \mathfrak{R}$.

1.2. Check the facts asserted in Examples (1), (4), (5), and (6) of rings.

1.3. Prove Proposition 1.5.

1.4. Prove that every ideal in $\mathbf{Z}$ is principal.

1.5. Prove that $(2, x)$ is a maximal ideal in $\mathbf{Z}[x]$.

1.6. Let R be a commutative ring, $b \in R$, and let $e_b: R[x] \to R$ be the substitution function that associates with each polynomial $p = a_0 +$

$a_1 x + \cdots + a_n x^n$ the element $a_0 + a_1 b + \cdots + a_n b^n$ of R. Show that e_b is a homomorphism and identify its kernel.

1.7. Prove the fact stated in Example (4) of an ideal.

1.8. Generalize the fact stated in Example (6) of a two-sided ideal.

1.9. Show that $R[G] \cong R[G]^0$ if R is commutative.

1.10. Show that, in the category of unitary rings, any two objects have a product.

2.1. Prove Proposition 2.2.

2.2. Define the *center* $Z(R)$ of the ring R as the set of those elements $z \in R$ such that $zr = rz$ for all $r \in R$. Show that $Z(R)$ is a commutative subring of R. Is it an ideal? Show that $\mathrm{Hom}_R(A, B)$ may be given the structure of a $Z(R)$-module.

2.3. Prove Theorem 2.11.

2.4. Give an example of R-modules A, B and an element ϕ of $\mathrm{Hom}_z(A,B)$ that does not belong to $\mathrm{Hom}_R(A, B)$.

2.5. (*Five lemma*) Consider the commutative diagram, with exact rows,

$$
\begin{array}{ccccccccc}
A_0 & \xrightarrow{\alpha_0} & A_1 & \xrightarrow{\alpha_1} & A_2 & \xrightarrow{\alpha_2} & A_3 & \xrightarrow{\alpha_3} & A_4 \\
\downarrow{\scriptstyle \phi_0} & & \downarrow{\scriptstyle \phi_1} & & \downarrow{\scriptstyle \phi_2} & & \downarrow{\scriptstyle \phi_3} & & \downarrow{\scriptstyle \phi_4} \\
B_0 & \xrightarrow{\beta_0} & B_1 & \xrightarrow{\beta_1} & B_2 & \xrightarrow{\beta_2} & B_3 & \xrightarrow{\beta_3} & B_4
\end{array}
$$

in $_R\mathfrak{M}$. Show that if $\phi_0, \phi_1, \phi_3, \phi_4$ are isomorphisms, so is ϕ_2. Can the hypotheses be weakened without sacrificing the conclusion?

2.6. Let

$$
\begin{array}{ccccc}
A' & \xrightarrow{\mu_1} & A & \xrightarrow{\epsilon_1} & A'' \\
\downarrow{\scriptstyle \phi'} & & \downarrow{\scriptstyle \phi} & & \downarrow{\scriptstyle \phi''} \\
B' & \xrightarrow{\mu_2} & B & \xrightarrow{\epsilon_2} & B''
\end{array}
$$

be a commutative diagram, with exact rows, in $_R\mathfrak{M}$. Show that, given ϕ' and ϕ'', ϕ is determined *modulo* an arbitrary homomorphism $A'' \to B'$. Given ϕ', ϕ'' can one always find ϕ to make the diagram commute?

3.1. Show that (3.2) gives $\mathrm{Hom}_R(A, B)$ the structure of a right T-module.

3.2. Let $\omega: R \to S$ be a ring homomorphism. If A is an S-module, regard it as an R-module by the rule

$$ra = \omega(r)a, \quad r \in R, \quad a \in A.$$

Let A^ω be the resulting R-module. Show that $A \mapsto A^\omega$ yields a functor $U^\omega: {}_S\mathfrak{M} \to {}_R\mathfrak{M}$. Describe the dependence of U^ω on ω.

3.3. Let $F^\omega: {}_R\mathfrak{M} \to {}_S\mathfrak{M}$ be given by $F^\omega B = \operatorname{Hom}_R(S^\omega, B)$. Here ω is as in Exercise 3.2 and S^ω is given the structure of an (R, S)-bimodule so that, according to Theorem 3.1, $F^\omega B$ has the structure of an S-module. Show that U^ω is left adjoint to F^ω.

3.4. Show that there is a natural homomorphism

$$\oplus \operatorname{Hom}_R(A, B_i) \to \operatorname{Hom}_R(A, \oplus B_i).$$

Is it an isomorphism? Is it a monomorphism? Is it an epimorphism?

3.5. Make a study, similar to that in Exercise 3.4, of

$$\oplus \operatorname{Hom}_R(A_i, B) \to \operatorname{Hom}_R(\Pi A_i, B).$$

4.1. Prove Proposition 4.1.

4.2. Let P be a family of primes (in $\mathbf{Z}$) and $\mathbf{Z}_{(P)}$ be the subring of $\mathbf{Q}$ consisting of rational numbers expressible as fractions a/b, where

$$p \in P \Rightarrow p \nmid b.$$

Show that the $\mathbf{Z}_{(P)}$-modules are just the tensor products $A \otimes \mathbf{Z}_{(P)}$, where A is an abelian group. Show that, if A, B are $\mathbf{Z}_{(P)}$-modules then

$$\operatorname{Hom}_{\mathbf{Z}}(A, B) = \operatorname{Hom}_{\mathbf{Z}_{(P)}}(A, B), \quad A \otimes_{\mathbf{Z}} B = A \otimes_{\mathbf{Z}_{(P)}} B.$$

4.3. Complete the proof of Theorem 4.3.

4.4. Show that the functor $A \otimes_S -$ is not, in general, exact.

4.5. Show that the functor $A \otimes_S -$ is right exact without using Theorem 4.3.

4.6. Set up a natural transformation $\eta: \operatorname{Hom}_R(A, R) \otimes_R B \to \operatorname{Hom}_R(A, B)$ of functors

$$_R\mathfrak{M}_S^0 \times {}_R\mathfrak{M}_T \to {}_S\mathfrak{M}_T.$$

Show that η is an isomorphism when A or B is a finite direct sum of copies of R and $S = T = \mathbf{Z}$.

4.7. Show that the functor $L^\omega: {}_R\mathfrak{M} \to {}_S\mathfrak{M}$, given by $L^\omega B = S^\omega \otimes_R B$, is left adjoint to $U^\omega: {}_S\mathfrak{M} \to {}_R\mathfrak{M}$ (see Exercises 3.2 and 3.3).

5.1. Let $U: {}_R\mathfrak{M} \to \mathfrak{S}$ be the underlying functor to sets. Show that the *free R-module functor* is left adjoint to U.

5.2. Show that every vector space has a basis (if the vector space is not finitely generated, you will need Zorn's lemma).

5.3. Show that $\mathbf{Q}$ is not a projective abelian group.

5.4. Consider the functor $U: {}_S\mathfrak{M} \to {}_R\mathfrak{M}$ of Exercise 3.2. Show that, if S^ω

is a projective R-module, then U^ω maps projective modules to projective modules, without appeal to the right adjoint functor F^ω.

5.5. In the notation of Theorem 5.14, show that θ_{A-} is a natural equivalence if and only if θ_{AA} is surjective.

6.1. Show that 0 is the only abelian group that is both projective and injective.

6.2. Show that a direct sum of injective abelian groups is injective.

6.3. Show that $\mathbf{Q}$ may be embedded as a direct factor in a direct product of copies of $\mathbf{Q}_1$.

6.4. Show that $\mathrm{Hom}(\mathbf{Q},\mathbf{Q}) \cong \mathbf{Q}$. (Harder) Show that $\mathrm{Hom}(\mathbf{Q},\mathbf{Q}_1)$ has the power of the continuum.

6.5. Show that the set of R-homomorphisms $A \to B$ that factor through an injective module form a subgroup of $\mathrm{Hom}_R(A,B)$. State and prove the "dual."

6.6. Show that an R-module A is injective if and only if, for any left ideal J in R, every homomorphism $J \to A$ extends to R.

6.7. Consider the sequence

$$\sum : A' \to A \to A''$$

in $_R\mathfrak{M}$ and the induced sequence

$$\sum^{\ddagger} : A''^{\ddagger} \to A^{\ddagger} \to A'^{\ddagger}$$

in $\mathfrak{M}_R$, where $B^{\ddagger} = \mathrm{Hom}(B,\mathbf{Q}_1)$. Show that $\sum$ is short exact if and only if $\sum^{\ddagger}$ is short exact.

5

INTEGRAL DOMAINS

1. PRINCIPAL IDEAL DOMAINS

We call an integral domain R a *principal ideal domain* (pid) if every ideal in R is principal. Thus every field is trivially a pid; the ring $\mathbf{Z}$ is also a pid, since we know that the ideals of $\mathbf{Z}$ are merely the sets of multiples of a fixed integer n. We will proceed quickly to establish the existence of other pid's by means of the following supplementary definition.

A *Euclidean function* on an integral domain R is a function $d: R \to \mathbf{Z}^{+}$, the nonnegative integers, satisfying the conditions

$E_1: dr = 0 \Leftrightarrow r = 0.$

$E_2: dr \leqslant d(rs)$ if $s \neq 0$, and $d(1) = 1.$

$E_3:$ Given $a, b \in R, b \neq 0$, there exist $q, r \in R$ with $a = bq + r$, and $dr < db.$
Of course, axiom E_3 is the crucial condition justifying the name of the function. We will often find d to be multiplicative, that is,

$$d(rs) = (dr)(ds), \quad r, s \in R. \tag{1.1}$$

Plainly, axiom E_2 follows from (1.1) and axiom E_1.

A *Euclidean domain* is an integral domain admitting a Euclidean function.

PROPOSITION 1.1. *Let* d *be a Euclidean function on the integral domain* R. *Then*

(1) $dr = 1 \Leftrightarrow r$ *is a unit.*

(*ii*) *If* r *is nonzero,* $dr = d(rs) \Leftrightarrow s$ *is a unit.*

PROOF. Since $d1 = 1$, (i) is implied by (ii), so we prove (ii). First let s be a unit. Then there exists t with $st = 1$, so that $r = rst$ and $dr \leqslant d(rs)$, $d(rs) \leqslant dr$ by axiom E_2.

Conversely, suppose that $dr = d(rs)$. Then, since r is nonzero, s is nonzero by axiom E_1. We prove that, if s is a nonunit, then $dr < d(rs)$. For

159

if s is a nonunit, then $rs \nmid r$. Thus, by axiom E_3, since $rs \neq 0$,

$$r = rst + u, \quad du < d(rs),$$

and $u \neq 0$. Now $u = r(1 - st)$. Thus, by axiom E_2, $dr \leq du$, whence $dr < d(rs)$.

THEOREM 1.2. *A Euclidean domain is a pid.*

PROOF. Let d be a Euclidean function on the integral domain R and let I be an ideal in R. If $I = (0)$ it is certainly principal, so we suppose $I \neq (0)$. Then $dI \subseteq \mathbf{Z}^+$ contains nonzero integers, hence a smallest positive integer d_0. Let $db = d_0, b \in I$; we prove that $I = (b)$. Certainly $(b) \subseteq I$. Conversely, let $a \in I$. Since $b \neq 0$, we may find $q, r \in R$ with $a = bq + r$, $dr < db$. But then $r = a - bq \in I$. By the minimality of db, we infer that $dr = 0$, so that $r = 0$. Thus $a \in (b)$ and $I = (b)$.

We may thus find pid's by discovering Euclidean domains.

EXAMPLES.

1. $\mathbf{Z}$ is a Euclidean domain, with $dr = |r|$.

2. Let $F[x]$ be the integral domain of polynomials in the indeterminate x over the field F. We define $d: F[x] \to \mathbf{Z}^+$ by $d(p) = 0, p = 0$ and $d(p) = 2^{deg p}, p \neq 0$. Then the usual division algorithm yields axiom E_3.

3. Let $\mathbf{Z}_{(p)}$ be the subring of Q consisting of rational numbers expressible as fractions m/n with $p \nmid n$. We define

$$d: \mathbf{Z}_{(p)} \to \mathbf{Z}^+ \text{ by } d(0) = 0,$$

$$d\left(\frac{m}{n}\right) = p^q, \quad \text{where } m = p^q m', p \nmid m'.$$

4. Let Γ be the ring of *Gaussian integers*; that is, Γ is the subring of the field $\mathbf{C}$ of complex numbers, consisting of complex numbers $a + ib$ with $a, b \in \mathbf{Z}$. We define $d: \Gamma \to \mathbf{Z}^+$ by

$$d(a + ib) = a^2 + b^2.$$

We leave the reader to verify axiom E_3 in Examples (3) and (4); and to show that, in all these examples, (1.1) is satisfied.

5. We give a counterexample. We know that $\mathbf{Z}[x]$ is not a pid, since the ideal $(2, x)$ is not principal. Of course, it is clear that the division algorithm cannot be executed in $\mathbf{Z}[x]$, but this says more; it says that $\mathbf{Z}[x]$ admits *no* Euclidean function. Note that the function d of Example (2) does not satisfy Proposition 1.1(i) on $\mathbf{Z}[x]$.

Let R be an integral domain and let S be a subset of R containing a nonzero element. We define the *gcd (greatest common divisor)* of S in R to

be an element $d \in R$, if such exists, such that
(i) $d|a$ for a in S.
(ii) If $r|a$ for all a in $S, r \in R$, then $r|d$.

PROPOSITION 1.3.
The gcd of S, if it exists, is unique up to multiplication by a unit.
We leave the proof to the reader.

THEOREM 1.4. *Let* R *be a pid, and* S *be a subset of* R *containing a nonzero element. Let* I *be the ideal generated by* S, *so that* I = (d), *for some element* d ∈ R. *Then* d *is the gcd of* S *in* R.

PROOF. Let $I = (S) = (d)$. Then if $a \in S, a \in (d)$, so that $d|a$. Next $d \in I$, so that d may be written

$$d = \sum r_i a_i \text{ (finite sum)}, \quad r_i \in R, \quad a_i \in S \qquad (1.2)$$

[see (1.2) of Chapter 4]. It follows immediately from (1.2) that if $r|a$ for all a in S, then $r|d$. Relation (1.2) expresses a particularly important fact about gcd's in a pid; *the gcd of* S *is a linear combination of elements of* S.

We now pursue further divisibility questions in a pid. However, our definitions will be more general in order to prepare for the next section. If R is an integral domain, we say that $a \in R$ is *irreducible* if a is not itself a unit, but an expression of a as $a = bc$ requires that b or c be a unit. Thus in **Z** the irreducible elements are precisely the prime numbers. Now we gave a definition of a prime ideal in Section 1 of Chapter 4; we say that $a \in R$ is *prime* if (a) is a prime ideal. Recall that 0 is *not* a prime ideal, according to our convention. We now prove a theorem that generalizes our experience of **Z**.

THEOREM 1.5. *Let* R *be a pid and let* a ∈ R. *Then the following statements are equivalent.*
(i) a *is prime.*
(ii) a *is irreducible.*
(iii)(a) *is maximal nonzero.*

PROOF
(i)⟹(ii) Let a be prime and $a = bc$. Then $a|bc$, so that, a being prime, $a|b$ or $a|c$. Suppose that $a|b$. Since also $b|a$, it follows that a and b are associated, so that c is a unit. (Notice that this argument is valid in any integral domain.)
(ii)⟹(iii) Let a be irreducible and let (a) be properly contained in the ideal (b). We will show that $(b) = R$. Since $(a) \subset (b), a = bc$ for some $c \in R$. Since a is irreducible, b or c is a unit. Were c a unit, we would have $b|a$, so

that $(a) = (b)$, contrary to hypothesis. Thus b is a unit, that is, $(b) = R$.

(iii)$\Rightarrow$(i) Theorem 1.7 of Chapter 4.

We are now ready to prove the fundamental theorem of unique factorization in a pid R. A *decomposition* of an element $a \in R$ (where R is any integral domain) is an expression for a as

$$a = up_1 \cdots p_k, \tag{1.3}$$

where u is a unit and each p_i is irreducible. We say that the decomposition is *unique* if, whenever we have two decompositions,

$$up_1 \cdots p_k = vq_1 \cdots q_l,$$

then $k = l$ and, renumbering the q's if necessary, p_i and q_i are associated, $i = 1, 2, \ldots, k$. We say that R is a *unique factorization domain* (ufd) if every nonzero element of R has a unique decomposition.

THEOREM 1.6. *A pid is a ufd.*

PROOF. We first require a lemma, whose conclusion holds in far more general circumstances (see Theorem 3.1).

LEMMA 1.7. *Let* $\cdots \subseteq I_n \subseteq I_{n+1} \subseteq \cdots$ *be an ascending chain of ideals in the pid R. Then there exists* m *such that* $I_m = I_{m+1} = \cdots$.

PROOF OF LEMMA. Let $I = \cup_n I_n$. It is easy to see that I is an ideal. Since R is a pid, $I = (a)$, for some $a \in R$. Since $a \in I, a \in I_m$ for some m. But then $(a) \subseteq I_m$, so that $I = I_m$, and $I_m = I_{m+1} = \cdots$. We express the conclusion of this lemma by saying that R satisfies the *ascending chain condition* on ideals.

We now return to the proof of the theorem. We first prove the existence of a decomposition. Thus let us suppose that there is an element $a_0 \in R$ which has no decomposition. Then a_0 cannot be irreducible, for, if so, $a_0 = 1a_0$ would be a decomposition. Thus $a_0 = a_1 b_1$ with neither a_1 nor b_1 a unit. Moreover, at least one of a_1, b_1 must be indecomposable. For if $a_1 = u_1 p_1 \cdots p_k$, $b_1 = u_2 q_1 \cdots q_l$ were decompositions, then $a_1 b_1 = u_1 u_2 p_1 \cdots q_l$ would be a decomposition of a_0. Suppose then that a_1 admits no decomposition. Since b_1 is not a unit, we have (a_0) properly contained in (a_1),

$$(a_0) \subset (a_1).$$

We now repeat the argument with a_1 replacing a_0. We find in this way a chain of ideals

$$(a_0) \subset (a_1) \subset (a_2) \subset \cdots,$$

each properly contained in its successor. But this is impossible by the lemma.

We now prove the uniqueness. Suppose that

$$up_1 \cdots p_k = vq_1 \cdots q_l; \qquad (1.4)$$

we argue by induction on k. If $k = 0$, then $vq_1 \cdots q_l$ is a unit. But it is obvious that if a product of elements of R is a unit, each is a unit. Thus $l = 0$ as required. Returning to (1.4) we have

$$p_k | vq_1 \cdots q_l.$$

Now p_k, being irreducible, is prime by Theorem 1.5. Thus (by an obvious induction!) $p_k | v$, or $p_k | q_1$, or..., or $p_k | q_l$. It is impossible that $p_k | v$, since p_k is not a unit; so, renumbering if necessary, we may suppose that $p_k | q_l, q_l = p_k w$. But q_l is irreducible and p_k is not a unit, so that w is a unit. We thus infer from (1.4) that p_k, q_l are associated and

$$up_1 \cdots p_{k-1} = vwq_1 \cdots q_{l-1}.$$

Our inductive hypothesis now comes into play to enable us to deduce that $k - 1 = l - 1$ and, renumbering the q's if necessary, p_i is associated with $q_i, i = 1, 2, \ldots, k - 1$. Thus $k = l, p_k$ is associated with q_k, and uniqueness is proved.

It may readily be shown that, in any ufd, the standard elementary theory of gcd's and lcm's (least common multiples) may be applied. Thus it is important to identify other ufd's apart from pid's—for example, we would like to show that $\mathbf{Z}[x]$ is a ufd. The main theorem of the next section will achieve this.

2. UNIQUE FACTORIZATION DOMAINS

We established in the preceding section that a principal ideal domain (pid) is a unique factorization domain (ufd). Our object here is to show how we may use this fact to identify other ufd's—we should know why $\mathbf{Z}[x]$ is a ufd —and we want to show how much of the elementary arithmetic theory of pid's passes to ufd's. We first discuss this second question.

THEOREM 2.1. *Let* R *be a ufd and let* $a \in R$. *Then* a *is irreducible if and only if* a *is prime.*

PROOF. We saw in the proof of Theorem 1.5 that in any integral domain a prime element is irreducible. Suppose, conversely, that R is an ufd and that $a \in R$ is irreducible. Suppose further that $a | bc$. Thus there exists $r \in R$ with $ar = bc$. Let us decompose r, b, c as

$$r = u\pi_1\pi_2 \cdots \pi_j, \quad u \text{ unit}, \quad \pi_i \text{ irreducible},$$

$$b = vp_1p_2 \cdots p_k, \quad v \text{ unit}, \quad p_i \text{ irreducible},$$

$$c = wq_1q_2 \cdots q_l, \quad w \text{ unit}, \quad q_i \text{ irreducible}.$$

Then $ua\pi_1\pi_2 \cdots \pi_j = vwp_1p_2 \cdots p_kq_1q_2 \cdots q_l$. These are both decompositions, since a is irreducible. Thus, by the uniqueness of the decomposition, a is associated with some p_i or some q_i—in the former case, $a|b$; in the latter, $a|c$.

The principle that we have used in this argument may be stated as follows.

PROPOSITION 2.2. *Let* R *be a ufd and let* a, b $\in$ R $- (0)$. *Then* a$|$b *if and only if, given a decomposition of* b, *there exists a decomposition of* a *whose constituent irreducibles form a subset of the irreducibles in the decomposition of* b.

A formal proof may be left to the reader. Notice that, in general, the irreducibles constituting the decomposition of an element of $R - (0)$ occur with repetitions.

THEOREM 2.3. *Let* R *be a ufd and let* S *be a subset of* R *containing a nonzero element. Then* S *has a gcd.*

PROOF. Let P be a collection of irreducibles of R containing precisely one irreducible from each equivalence class of associated irreducibles. We will agree, in writing a decomposition of an element of R, only to employ irreducibles from P. Then, since R is a ufd, we may write each nonzero element of R in a unique way as

$$a = up_1p_2 \cdots p_k, \quad u \text{ unit}, \quad p_i \in P,$$

and the rule $a \mapsto (p_1, p_2, \ldots, p_k)$ yields a function

$$f : R - (0) \to \text{Fin}(P),$$

where $\text{Fin}(P)$ stands for the set of finite subsets of P *with repetitions allowed.*

Now, given S, we restrict f to $S - (0)$ which is nonempty by hypothesis. Then

$$D = \bigcap_{s \in S - (0)} f(s)$$

is an element of $\mathrm{Fin}(P)$. With any element E of $\mathrm{Fin}(P)$ we may associate an element

$$d(E) = \prod_{p \in E} p$$

of R, such that $fd(E) = E$. We claim that $d(D)$ is the gcd of S.

For Proposition 2.2 asserts that $a|b$ if and only if $f(a) \subseteq f(b)$. Since $fd(D) \subseteq f(s)$ for every $s \in S - (0)$, we have $d(D)|s$ for every $s \in S - (0)$. Of course $d(D)|0$, so that $d(D)|s$ for every $s \in S$. If $r|s$ for every $s \in S$ then $r \neq 0$ and $f(r) \subseteq f(s)$ for every $s \in S - (0)$, so that $f(r) \subseteq fd(D)$ and $r|d(D)$.

The reader should surely recognize this argument as a mild generalization of the traditional "schoolboy" process of finding gcd's.

We now proceed to enunciate the main theorem of this section. The proof will require some preparation but we state the theorem immediately to provide motivation.

THEOREM 2.4. *If R is a ufd, so is the polynomial ring* R[x].

COROLLARY 2.5. *If R is a pid,* R[x] *is a ufd.*

This explains, in particular, why $\mathbf{Z}[x]$ is a ufd.

COROLLARY 2.6. *If F is a field, then* F[$x_1, x_2, \ldots, x_n$] *is a ufd.*

Indeed $F[x]$ is a pid. It is now only necessary to observe that, for any integral domain R

$$R[x_1, x_2, \ldots, x_n] = R[x_1, x_2, \ldots, x_{n-1}][x_n], \tag{2.1}$$

so that we may argue by induction. Of course, Corollary 2.6 is only a special case of

COROLLARY 2.7. *If R is a ufd, so is* R[$x_1, x_2, \ldots, x_n$].

However, we preferred to enunciate Corollary 2.6 explicitly, since $F[x]$ is more than a ufd.

We now prepare for the proof of Theorem 2.4. First we make a universal construction in the category of integral domains.

DEFINITION 2.1. Let R be an integral domain. By the *quotient field* of R we understand a pair consisting of a field F and a ring injection $i : R \to F$, enjoying the following universal property. Given any field Φ and injection $f : R \to \Phi$, there exists a unique homomorphism $g : F \to \Phi$ with $gi = f$.

We now prove the existence of the quotient field; as always with such universal constructions, uniqueness follows from the definition. The construction we give of F is based on the familiar construction of **Q** out of **Z**. We construct the set V of pairs (r,s), $r,s \in R, s \neq 0$, and introduce into V the equivalence relation

$$(r_1, s_1) \sim (r_2, s_2) \qquad \text{if and only if} \quad r_1 s_2 = r_2 s_1.$$

It is easy to see that this is an equivalence relation. Let F be the set of equivalence classes; we write $[r,s]$ for the equivalence class of (r,s). We define addition and multiplication in F by the rules

$$[r,s] + [r',s'] = [rs' + r's, ss'],$$

$$[r,s][r',s'] = [rr', ss'].$$

We check that these rules are well-defined and that F thereby becomes a commutative ring with $0 = [0,1]$, $1 = [1,1]$. Indeed, F becomes a field. For if $[r,s] \neq 0$, then $r \neq 0$, so that $[s,r] \in F$ and

$$[r,s][s,r] = [rs, rs] = [1,1] = 1.$$

We define $i : R \to F$ by

$$i(r) = [r,1].$$

Then i is obviously a homomorphism; indeed it is monic, since, if $i(r) = 0$, then $[r,1] = [0,1], r = 0$.

THEOREM 2.8. *The pair* (F,i) *constructed above is the quotient field of* R.

PROOF. Suppose that the injection $f : R \to \Phi$ is given. We define $g : F \to \Phi$ by

$$g[r,s] = f(r)f(s)^{-1}.$$

This is well-defined, since $s \neq 0$, so that $f(s) \neq 0$, and plainly $gi = f$. One easily checks that g is a homomorphism. Moreover, g is obviously the unique homomorphism from F to Φ with $gi = f$. For if g is a homomorphism with $gi = f$, then

$$g[r,s] = g\big([r,1][s,1]^{-1}\big) = g[r,1]\big(g[s,1]\big)^{-1} = fr(fs)^{-1}$$

We will adopt the notation familiar in **Q** and write r/s for the element $[r,s]$ of F represented by (r,s) in V. Now if R is a ufd, we may construct the gcd d of r and s, and we have $r = r_0 d$, $s = s_0 d$. Moreover, it is plain that the gcd of r_0 and s_0 is 1. Thus we may always choose, as a representative *fraction* for an element of F, a fraction r_0/s_0 such that the gcd of r_0 and s_0 is 1. We say that r_0/s_0 is then *in lowest terms*.

PROPOSITION 2.9. *The embedding* $i : R \to F$ *extends to an embedding* $R[x] \subseteq F[x]$.

Thus we will henceforth suppose R embedded in F (by means of i) and $R[x]$ embedded in $F[x]$.

PROPOSITION 2.10. *If a is irreducible in* R *it is irreducible in* R[x].

We leave the proof to the reader.

We assume henceforth that R is a ufd. Then, given any nonzero polynomial

$$f = a_0 + a_1 x_1 + \cdots + a_n x^n$$

in $R[x]$, we may form $d = \gcd(a_0, a_1, \ldots, a_n), d \in R$. We call d the *height* of f and say that f is *monic* if its height is 1. Plainly if f has height d, then f has a unique expression

$$f = dg,$$

with g monic—and conversely.

PROPOSITION 2.11. *Let* f,g *be monic in* R[x]. *Then* fg *is also monic.*

PROOF. Let $f = a_0 + a_1 x + \cdots + a_m x^m, g = b_0 + b_1 x + \cdots + b_n x^n$ be monic. Then

$$fg = c_0 + c_1 x + \cdots + c_k x^k + \cdots,$$

where

$$c_k = \sum_{i+j=k} a_i b_j.$$

(As usual, we adopt the convention that $a_i = 0, i > m, b_j = 0, j > n$.) Suppose that fg is not monic. Then there must be some irreducible p, such that $p | c_k$, all k. Since f is monic, there exists i_0 such that

$$p | a_i, \quad i < i_0, \quad p \nmid a_{i_0}.$$

Similarly, g being monic, there exists j_0 such that

$$p | b_j, \quad j < j_0, \quad p \nmid b_{j_0}.$$

Let $i_0 + j_0 = k_0$ and consider c_{k_0}. Then

$$c_{k_0} = (a_0 b_{k_0} + \cdots + a_{i_0-1} b_{j_0+1}) + a_{i_0} b_{j_0} + (a_{i_0+1} b_{j_0-1} + \cdots + a_{k_0} b_0)$$

$$= r + a_{i_0} b_{j_0} + s.$$

Since $p|a_i, i < i_0, p|r$. Since $p|b_j, j < j_0, p|s$. But $p|c_{k_0}$, so that $p|a_{i_0}b_{j_0}$. However p is irreducible, hence prime (Theorem 2.1), so that $p|a_{i_0}$ or $p|b_{j_0}$. We have thus arrived at a contradiction, so that fg is, as claimed, monic.

PROPOSITION 2.12. *Let* $f \in F[x], f \neq 0$. *Then* f *may be expressed as*

$$f = \frac{r}{s}f', \tag{2.2}$$

where f' *is monic in* $R[x]$. *Moreover, this expression is essentially unique.*

PROOF.   We have $f = \alpha_0 + \alpha_1 x + \cdots + \alpha_n x^n, \alpha_i \in F$, so that

$$\alpha_i = \frac{a_i}{b_i}.$$

Set $s = b_0 b_1 \cdots b_n$. Then

$$f = \frac{1}{s}(c_0 + c_1 x + \cdots + c_n x^n), \quad c_i \in R.$$

Let $r = \gcd(c_0, c_1, \ldots, c_n)$. Then

$$f = \frac{r}{s}f',$$

with f' monic.

If $(r_1/s_1)g_1 = (r_2/s_2)g_2$, with g_1, g_2 monic, then $r_1 s_2 g_1 = r_2 s_1 g_2$. Comparing heights, we find that $r_1 s_2 = u r_2 s_1$, $g_2 = u g_1$, for some unit u of R. This expresses the uniqueness of (2.2).

We note that the units of $R[x]$ are precisely the units of R. This enables us to establish our final result preliminary to the proof of Theorem 2.4.

PROPOSITION 2.13.   *Let* f *be a monic polynomial of* $R[x]$ *of positive degree. Then* f *is irreducible in* $R[x]$ *if and only if it is irreducible in* $F[x]$.

PROOF.   Let f factorize as $f = gh$ in $R[x]$ where neither g nor h is a unit. Were g, for example, of zero degree, it would be an element of R dividing the height of f, thus a unit. It follows that g and h are of positive degree, so that $f = gh$ is a factorization of f as a product of nonunits of $F[x]$.

Conversely, let f factorize as $f = gh$ in $F[x]$ where neither g nor h is a unit of $F[x]$, so that g, h are of positive degree. Use Proposition 2.12 to write

$$g = \frac{a}{b}g', \quad h = \frac{r}{s}h',$$

where g', h' are monic in $R[x]$. Then $bsf = arg'h'$. But $g'h'$ is monic by Proposition 2.11, so that $ar = bsu$ for some unit u of R and $f = ug'h'$ is a factorization in $R[x]$, with g', h' nonunits of $R[x]$.

Notice that it follows from Propositions 2.10 and 2.13 that the irreducibles of $R[x]$ are precisely the irreducibles of R together with the monic polynomials of $R[x]$ (of positive degree) which are irreducible in $F[x]$.

PROOF OF THEOREM 2.4. Let $f \in R[x], f \neq 0$. We write $f = df', d \in R$, f' monic. We regard f' as belonging to $F[x]$, where it has a decomposition (since $F[x]$ is a pid)

$$f' = \frac{a}{b} g_1 g_2 \cdots g_k, \qquad (2.3)$$

where each g_i is irreducible in $F[x]$. By Proposition 2.12 there is no real loss of generality in supposing each g_i monic in $R[x]$. But then, by Proposition 2.13, each g_i is irreducible in $R[x]$. Also we obtain from (2.3) that $bf' = ag_1 g_2 \cdots g_k$. Since $g_1 g_2 \cdots g_k$ is monic (by Proposition 2.11) it follows that $a = bu$, with u a unit of R, so that

$$f' = u g_1 g_2 \cdots g_k.$$

Thus $f = udg_1 g_2 \cdots g_k$. Decompose d as $p_1 p_2 \cdots p_m$, p_i irreducible in R. Then p_i is irreducible in $R[x]$, and

$$f = u p_1 p_2 \cdots p_m g_1 g_2 \cdots g_k$$

is a decomposition of f in $R[x]$.

To prove uniqueness, suppose that

$$u p_1 p_2 \cdots p_m g_1 g_2 \cdots g_k = v q_1 q_2 \cdots q_n h_1 h_2 \cdots h_l$$

(in the obvious notation). We pass to $F[x]$, where $u p_1 p_2 \cdots p_m$, $v q_1 q_2 \cdots q_n$ are units. Thus by the uniqueness of decomposition in $F[x]$ we infer that $k = l$ and, renumbering if necessary, each g_i is associated with h_i in $F[x]$. But if

$$g_i = \frac{r_i}{s_i} h_i,$$

we infer, as usual, from the fact that g_i and h_i are monic that $r_i/s_i = w_i$, a unit of R. Thus g_i, h_i are associated in $R[x]$ and

$$u p_1 p_2 \cdots p_m = w q_1 q_2 \cdots q_n,$$

where $w = v(w_1 w_2 \cdots w_k)^{-1}$. We now invoke the uniqueness of decomposition in R to infer that $m = n$, and, renumbering if necessary, each p_i is associated with q_i in R, hence in $R[x]$. This completes the proof.

We close this section by exhibiting an integral domain that is not a ufd.

Let R be the subset of the field of complex numbers $\mathbf{C}$, consisting of complex numbers $a + \sqrt{-5}\, b$, $a,b \in \mathbf{Z}$. It is very easy to see that R is a subring of $\mathbf{C}$, hence an integral domain. Let $d: R \to \mathbf{Z}^+$ be the function given by

$$d(a + \sqrt{-5}\, b) = a^2 + 5b^2.$$

Then d is just the restriction to R of the square of the absolute value of a complex number. Thus d is multiplicative [that is, it satisfies (1.1)]. It certainly satisfies axiom E_1—hence also axiom E_2—for a Euclidean function, but it will turn out that it fails to satisfy axiom E_3.

Now it is an immediate consequence of the fact that d is multiplicative that $du = 1$ if u is a unit. On the other hand, if $a,b \in \mathbf{Z}$, then $a^2 + 5b^2 = 1 \Leftrightarrow a = \pm 1, b = 0$. Since ± 1 are units of R, they are the only units. Consider now the equation

$$2 \cdot 3 = (1 + \sqrt{-5})(1 - \sqrt{-5}) \tag{2.4}$$

in R. We claim that $2, 3, 1 \pm \sqrt{-5}$ are all *irreducible* in R. We will be content to prove this for 2; the other three claims are treated similarly. Suppose then that $2 = \alpha\beta$ in R. Applying d, we have

$$4 = (d\alpha)(d\beta)$$

in $\mathbf{Z}^+$. If $d\alpha = 1$, then α is a unit; similarly for β. Thus if 2 were not irreducible we would have to admit $d\alpha = 2, d\beta = 2$. But d does not take the value 2 on R.

Thus (2.4) marks two factorizations of 6 as a product of irreducibles. However, since the only units of R are ± 1, it is plain that uniqueness is contradicted. We may observe that, in fact, 2, 3, and $1 \pm \sqrt{-5}$ are irreducible elements of R that are not prime.

The reader should observe that one easily proves, utilizing the function d and induction, that every element of R admits a decomposition as a product of irreducibles.

3. NOETHERIAN RINGS AND MODULES

In this section we generalize pid's in a different direction.

DEFINITION 3.1.   A left R-module A is *Noetherian* if every submodule of A is finitely generated. A ring R is left *Noetherian* if it is a Noetherian left R-module.

Thus R is left Noetherian if and only if all its left ideals are finitely generated. Thus a pid is certainly left (and right) Noetherian. We will see later that $Z[x]$ is Noetherian, but, of course, it is not a pid.

Recalling Lemma 1.7 we prove

THEOREM 3.1. *Let* $A \in {}_R\mathfrak{M}$. *Then the following statements about* A *are equivalent*:

(*i*) A *is Noetherian.*

(*ii*) A *satisfies the ascending chain condition.*

(*iii*) *Every nonempty collection of submodules of* A *has a maximal element.*

PROOF

(i)⇒(ii) Clearly (ii) claims that if $\cdots \subseteq A_n \subseteq A_{n+1} \subseteq \cdots$ is an ascending chain of submodules of A, then there exists m such that $A_m = A_{m+1} = \cdots$. We prove this just as we proved Lemma 1.7. Consider

$$A' = \bigcup_n A_n.$$

Plainly A' is a submodule of A, hence, A being Noetherian, finitely generated. Let $(b_1, b_2, \ldots, b_k)$ be a set of generators of A'. Then $b_i \in A_{n_i}$ for some n_i. Let $m = \max(n_1, n_2, \ldots, n_k)$. Then $b_i \in A_m, i = 1, 2, \ldots, k$, so that $A' \subseteq A_m$, whence $A' = A_m$ and $A_m = A_{m+1} = \cdots$.

(ii)⇒(iii) Let $\mathfrak{F}$ be a nonempty collection of submodules of A without maximal element. Then plainly we can extract from $\mathfrak{F}$ a strictly increasing sequence of submodules of A, contradicting (ii).

(iii)⇒(i) Let $A' \subseteq A$ and let $\mathfrak{F}$ be the collection of finitely generated submodules of A'. Then $(0) \in \mathfrak{F}$, so that $\mathfrak{F}$ is not empty. Let B be a maximal element. Were B strictly contained in A' we could obviously enlarge B to a bigger finitely generated submodule of A', contradicting the maximality of B. Thus $B = A'$ and A' is finitely generated.

We next show that the Noetherian R-modules form a nice abelian subcategory of ${}_R\mathfrak{M}$. Indeed, we have only to prove the following theorem.

THEOREM 3.2. *Let* $0 \to A' \to A \xrightarrow{\epsilon} A'' \to 0$ *be an exact sequence in* ${}_R\mathfrak{M}$. *Then* A *is Noetherian if and only if* A' *and* A'' *are Noetherian.*

PROOF. Let A be Noetherian. It is then trivial from the definition that A' is Noetherian. Now let B be a submodule of A''. Then $\epsilon^{-1}B$ is a submodule of A, hence finitely generated. Since $B = \epsilon(\epsilon^{-1}B)$, B is also finitely generated, so that A'' is Noetherian.

Conversely, suppose A', A'' Noetherian, and let A_0 be a submodule of A. Then the sequence

$$0 \to A' \cap A_0 \to A_0 \xrightarrow{\epsilon_0} \epsilon A_0 \to 0,$$

where we regard A' as a submodule of A and $\epsilon_0 = \epsilon | A_0$, is obviously exact. Since $A' \cap A_0$ is a submodule of A', it is finitely generated; since ϵA_0 is a submodule of A'', it is finitely generated. It is now trivial to conclude that A_0 is finitely generated, so that A is Noetherian.

COROLLARY 3.3. *If* A_1, A_2 *are submodules of* A, *then* $A_1 + A_2$ *is Noetherian if and only if* A_1 *and* A_2 *are Noetherian.*

PROOF. Obviously A_1 and A_2 are Noetherian if $A_1 + A_2$ is Noetherian. Conversely, suppose that A_1, A_2 is Noetherian. Then $A_1 \oplus A_2$ is Noetherian by Theorem 3.2. But $A_1 + A_2$ is a quotient of $A_1 \oplus A_2$ and is therefore Noetherian, also by Theorem 3.2.

COROLLARY 3.4. *Let* R *be a left Noetherian ring. Then every finitely generated* R-module *is Noetherian.*

PROOF. A finitely generated R-module is a quotient of a *finite* direct sum of copies of R. But, by Theorem 3.2, such a finite direct sum is Noetherian and so also is any quotient of it.

We note

THEOREM 3.5. *Every finitely generated flat module over a left Noetherian ring is projective.*

PROOF. It is plain that if $F_1 \rightarrow F_0 \overset{\epsilon}{\rightarrow\!\!\!\rightarrow} A$ is a free presentation of A with F_0 finitely generated, then, R being left Noetherian, F_0 is Noetherian, so that $\ker \epsilon$ is finitely generated. Thus we may take F_1 to be finitely generated. This shows that a finitely generated module over ' a left Noetherian ring is finitely presented and we apply Theorem 5.15 of Chapter 4.

We now prove a theorem analogous to Theorem 2.4 which enables us to enlarge our store of Noetherian rings.

THEOREM 3.6. *Let* R *be a commutative ring. Then* R[x] *is Noetherian if* R *is Noetherian.*

PROOF. We show that every ideal J in $R[x]$ is finitely generated. Let J_k be the subset of R consisting of the coefficients of x^k in polynomials in J of degree $\leqslant k$. Since J is an ideal in $R[x]$, it is plain that J_k is an ideal in R. Also, $J_k \subseteq J_{k+1}$, since, if $a_k \in J_k$, there exists a polynomial

$$b_0 + b_1 x + \cdots + b_{k-1} x^{k-1} + a_k x^k$$

in J; and then

$$b_0 x + b_1 x^2 + \cdots + b_{k-1} x^k + a_k x^{k+1}$$

also belongs to J, so that $a_k \in J_{k+1}$. Thus we have an ascending chain

$$J_0 \subseteq J_1 \subseteq \cdots$$

of ideals of R, so that, by Theorem 3.1, there exists m with $J_m = J_{m+1} = \cdots$.

Let $J_k = (a_{k1}, \ldots, a_{kn_k}), 0 \leqslant k \leqslant m$, and let f_{kj} be a polynomial of J, of degree $\leqslant k$, whose x^k-coefficient is a_{kj}. We claim that the polynomials $f_{kj}, 1 \leqslant j \leqslant n_k, 0 \leqslant k \leqslant m$, generate J. If not, suppose that they generate J' and let g be a polynomial of J of lowest degree outside J'. Let $\deg g = d$, and let c be the coefficient of x^d in g. Then we have expressions

$$c = \sum r_j a_{dj} \quad \text{if} \quad d \leqslant m,$$

or

$$c = \sum r_j a_{mj} \quad \text{if} \quad d > m.$$

In the former case, $g - \sum r_j f_{dj}$ has lower degree than g and thus lies in J', which would force g to be in J'. In the latter case, $g - \sum r_j x^{d-m} f_{mj}$ has lower degree than g and thus lies in J', again forcing g to lie in J'. Thus, in any case, we reach a contradiction, so that the theorem is proved.

COROLLARY 3.7. *If R is a commutative Noetherian ring, so is* $R[x_1, x_2, \ldots, x_n]$.

We close by exhibiting a commutative ring that is plainly not Noetherian, namely $R[x_1, x_2, \ldots, x_n, \ldots]$. For, in this ring, we have the properly ascending chain of ideals

$$(x_1) \subset (x_1, x_2) \subset (x_1, x_2, x_3) \subset \cdots.$$

On the other hand, we note that if R is a ufd, so is $R[x_1, x_2, \ldots, x_n, \ldots]$. For any element of $R[x_1, x_2, \ldots, x_n, \ldots]$ lies, in fact, in some $R[x_1, x_2, \ldots, x_n]$ and its decomposition in $R[x_1, \ldots, x_n]$ is precisely its decomposition in $R[x_1, x_2, \ldots, x_n, \ldots]$.

4. MODULES OVER PRINCIPAL IDEAL DOMAINS

The theory of modules over pid's very closely resembles that of abelian groups (which can, of course, be regarded as a special case). Thus where we have already described the theory in Chapter 2, we will give the corresponding enunciations and leave the reader to make the necessary small changes in the proofs.

Thus throughout this section we will be considering D-modules where D is a given pid.

THEOREM 4.1. *A submodule of a free* D-*module is free. Hence, a projective* D-*module is free.*

PROOF. See Theorem 3.4 of Chapter 2.

THEOREM 4.2. *A* D-*module is injective if and only if it is divisible.*

PROOF. See Theorem 3.10 of Chapter 2; of course, we call a D-module A *divisible* if, for any $a \in A$, and $d \neq 0 \in D$, there exists $b \in A$ with $db = a$.

Our main aim is to prove the fundamental theorem on the structure of finitely generated D-modules; in this case, we cannot refer to the corresponding theorem for abelian groups, since we deliberately postponed the proof of that theorem in the special case. The reader should keep the case $D = \mathbf{Z}$ in mind in studying what follows.

Now if A is a D-module and $a \in A$, we may consider the subset of D consisting of elements d such that $da = 0$. This subset is plainly an ideal in D, called the *annihilator ideal* of a. Since D is a pid, this ideal has a generator r which we call the *order* of a; however, if $r = 0$ it is customary to say that a has *infinite order*. If $r \neq 0$, we call a a *torsion element* of A. Plainly, the torsion elements form a submodule of A. For if $r_1 a_1 = 0, r_2 a_2 = 0, r_1 \neq 0, r_2 \neq 0$, then $r_1 r_2 (a_1 - a_2) = 0$ and $r_1 r_2 \neq 0$; and if $ra = 0, r \neq 0$, then $r(sa) = 0$. We call this the *torsion submodule*, T, of A, and say that A is a *torsion module* if $A = T$. We say that A is *torsion-free* if $T = 0$.

PROPOSITION 4.3. *If* A *is a* D-*module and* T *its torsion submodule, then* A/T *is torsion free.*

PROOF. Let $a + T \in A/T$ be a torsion element. Thus there exists $r \neq 0$ with $r(a + T) = 0$, that is, $ra \in T$. But then there exists $s \neq 0$ with $s(ra) = 0$. Since $sr \neq 0$, it follows that $a \in T$ and $a + T = 0$.

THEOREM 4.4. *A finitely generated torsion-free* D-*module is free.*

PROOF. Since D is a pid, it is Noetherian. Thus if F is a finitely generated torsion-free D-module, it is Noetherian and there exists, by Theorem 3.1, a maximal free submodule M of F (which is also finitely generated). Let $(x_1, x_2, \ldots, x_k)$ generate F. For each x_i there exists $r_i \neq 0 \in D$ such that $r_i x_i \in M$. For, if not, $M \oplus (x_i)$ would be a free submodule of F properly containing M. Let $r = r_1 r_2 \cdots r_k$ and define $\lambda_r : F \to M$ by $\lambda_r(x) = rx$. Then λ_r is injective, since F is torsion free. Thus λ_r embeds F in the free submodule M, so that, by Theorem 4.1, F is free.

THEOREM 4.5. *Let* F *be a free* D-*module and let it have a generating set consisting of* n *elements. Then it has a basis consisting of* $\leqslant$n *elements. Moreover, every basis of* F *has the same number of elements.*

PROOF. Let Φ be the quotient field of D. Then $\bar{F} = F \otimes_D \Phi$ is a Φ-vector space. Moreover, every generating set of F passes (under the map $a \mapsto a \otimes 1$) to a generating set of $\bar{F}$ and every basis of F passes to a basis of $\bar{F}$. Thus the theorem follows from well-known facts of linear algebra.

We note that no use is made in the last theorem of the fact that the integral domain D is a pid.

We call the number of elements in a basis for F the *rank* of F.

COROLLARY 4.6. *A finitely generated D-module is the direct sum of its torsion submodule and a free module of finite rank.*

PROOF. Let A be a finitely generated D-module. Then A/T is finitely generated, hence, by Proposition 4.3 and Theorem 4.4, free. Since A/T is free, the exact sequence $0 \to T \to A \to A/T \to 0$ splits. By Theorem 4.5, A/T has finite rank.

Plainly, the rank of A/T is an invariant of A and determines the isomorphism class of A/T. Thus, in classifying finitely generated D-modules, it now suffices to consider torsion modules. We first prove a theorem applying to arbitrary torsion modules over D (not necessarily finitely generated).

Let T be a torsion module over D and let p be a prime in D. Consider the set T_p of elements of T whose orders are powers of p. We claim that they form a submodule of T. For if $a, b \in T_p$, then $p^\alpha a = 0, p^\beta b = 0$, so that $p^\gamma(a - b) = 0$, where $\gamma = \max(\alpha, \beta)$, $p^\alpha(ra) = 0, r \in D$. Thus the order of $a - b$ divides p^γ, hence is itself a power of p (since p is irreducible in a ufd). Similarly the order of ra is a power of p.

THEOREM 4.7. *There is a natural equivalence*

$$T \cong \bigoplus_p T_p$$

(*Here, of course, it is to be understood that one chooses precisely one prime from each equivalence class of associated primes.*)

PROOF. The map we choose from $\oplus_p T_p$ to T is simply that which embeds T_p in T. Thus we could write the conclusion of the theorem as $T = \bigoplus_p T_p$. The naturality follows from the observation that a homomorphism of torsion modules $\phi : T \to \bar{T}$ must send T_p to $\bar{T}_p$, as the reader will readily prove.

We first show that every $a \in T$ is expressible as

$$a = \sum_p a_p, \quad a_p \in T_p \quad \text{(finite sum)}.$$

Let the order of a admit the decomposition

$$r = p_1^{\alpha_1} \cdots p_k^{\alpha_k},$$

and let $r_i = r/p_i^{\alpha_i}$. Then gcd $(r_1, r_2, \ldots, r_k) = 1$. Thus, since D is a pid, we conclude (1.2) that there are elements $s_i \in D, i = 1, 2, \ldots, k$, such that

$$1 = r_1 s_1 + r_2 s_2 + \cdots + r_k s_k,$$

so that

$$a = s_1 (r_1 a) + s_2 (r_2 a) + \cdots + s_k (r_k a). \tag{4.1}$$

Now it is evident that the order of $r_i a$ is $p_i^{\alpha_i}$, so that $r_i a \in T_{p_i}$, and therefore so does $s_i r_i a$. Thus (4.1) is the expression of a which we seek.

Second, we must show that $T_p \cap (\sum_{p' \neq p} T_{p'}) = (0)$. Let x belong to this intersection, so that $p^\alpha x = 0$ and $x = x_1 + x_2 + \cdots + x_n, x_i \in T_{p_i}, p_i \neq p,$ $i = 1, 2, \ldots, n$. If the order of x_i is $p_i^{\alpha_i}$, then $qx = 0$, where $q = p_1^{\alpha_1} p_2^{\alpha_2} \ldots p_n^{\alpha_n}$. Since D is a ufd, it follows that gcd $(p^\alpha, q) = 1$, and so, since D is a pid, $1 = rp^\alpha + sq, r, s \in D$. Thus $x = 1x = rp^\alpha x + sqx = 0$, and the theorem is proved.

PROPOSITION 4.8. *Let* S *generate* T *and let* r_a *be the order of* $a \in S$. *Let* P *be the collection of primes* p *such that* $p|r_a$ *for some* $a \in S$. *Then* $T_p = (0)$ *if and only if* $p \notin P$.

PROOF. Let $b \in T$, so that $b = \sum_{i=1}^k d_i a_i, \ d_i \in D, a_i \in S$. Let $r = r_{a_1} \cdots r_{a_k}$. Then $rb = 0$, so that the order of b divides r. But if $p \notin P$, then $p \nmid r$, so that the order of b cannot be a positive power of p, and $T_p = (0)$.

Conversely, let $p \in P$. Then there exists $a \in S$ whose order is pq, say, $q \in D$. Then the order of qa is p, so that $qa \neq 0, qa \in T_p$.

Theorem 4.7 and Proposition 4.8 show that, if T is a finitely generated torsion module over D, then T may be canonically expressed as a *finite* direct sum,

$$T = \bigoplus_{i=1}^k T_{p_i}, \tag{4.2}$$

and, of course, each T_{p_i} is finitely generated. Thus it finally remains only to consider a finitely generated torsion module $T = T_p$, all of whose elements have orders a power of a fixed prime p. We call a torsion module all of whose elements have orders a power of p a *p-module*, so that the next theorem completes the classification of finitely generated D-modules.

THEOREM 4.9. *Let* T $\neq (0)$ *be a finitely generated* p-module. *Then*

$$T \cong C_{p^{\alpha_1}} \oplus \cdots \oplus C_{p^{\alpha_k}}, \qquad \alpha_1 \geq \alpha_2 \geq \cdots \geq \alpha_k \geq 1,$$

where $C_{p^{\alpha_i}}$ is a cyclic module whose generator has order p^{α_i}. Moreover, the positive integers $\alpha_1, \alpha_2, \ldots, \alpha_k$ are uniquely determined by T.

PROOF. We consider all possible finite generating sets of T, and choose those with fewest elements, say k elements. With each such generating set we associate a k-tuple of positive integers, arranged in nonincreasing order, being the exponents of p in the orders of the elements of the generating set. Among the associated k-tuples we choose one, $(\alpha_1, \alpha_2, \ldots, \alpha_k)$, such that $\sum \alpha_i$ is minimal. Let $(a_1, a_2, \ldots, a_k)$ be the corresponding generating set, so that a_i has order $p^{\alpha_i}, \alpha_1 \geqslant \alpha_2 \geqslant \ldots \geqslant \alpha_k \geqslant 1$. We claim that this generating set yields the direct sum decomposition claimed in the theorem, that is, we claim that $a_1, a_2, \ldots, a_k$ are linearly independent over[†] D.

If not, there exists i, $0 \leqslant i \leqslant k-1$, such that $a_1, a_2, \ldots, a_i$ are linearly independent, but a_{i+1} is linearly dependent on $a_1, a_2, \ldots, a_i$. Let (d) be the ideal consisting of those elements $r \in D$ such that $ra_{i+1} \in (a_1, a_2, \ldots, a_i)$. Since $p^{\alpha_{i+1}}a_{i+1} = 0$, we have $d \mid p^{\alpha_{i+1}}$, so that $d = p^{\beta}, \beta < \alpha_{i+1}$, since, for some r, $0 \neq ra_{i+1} \in (a_1, a_2, \ldots, a_i)$. Now

$$p^{\beta}a_{i+1} = \sum_{j=1}^{i} \lambda_j p^{\gamma_j}a_j, \quad p \nmid \lambda_j,$$

so that

$$0 = p^{\alpha_{i+1}}a_{i+1} = \sum_{j=1}^{i} \lambda_j p^{\alpha_{i+1}-\beta+\gamma_j}a_j.$$

By our hypothesis, this forces $\alpha_{i+1} - \beta + \gamma_j \geqslant \alpha_j, 1 \leqslant j \leqslant i$, so that certainly $\gamma_j \geqslant \beta$. Set $a' = a_{i+1} - \sum_{j=1}^{i}\lambda_j p^{\gamma_j - \beta}a_j$. Then we get a new generating set by replacing a_{i+1} by a'. But $p^{\beta}a' = 0, \beta < \alpha_{i+1}$, contradicting the minimality of $\sum \alpha_i$.

Note that the role played by the minimality of k is to ensure that none of the α_i can be zero.

We now prove that the integers $\alpha_1, \alpha_2, \ldots, \alpha_k$ are uniquely determined. If $T \cong C_{p^{\alpha_1}} \oplus \cdots \oplus C_{p^{\alpha_k}}$ $\alpha_1 \geqslant \alpha_2 \geqslant \cdots \geqslant \alpha_k \geqslant 1$, it is plain that α_1 is characterized as the smallest integer n such that $p^n T = 0$. We call α_1 the p-exponent of T. We complete the argument by induction on α_1. Thus we suppose that

$$T \cong \bigoplus_{i=1}^{k} C_{p^{\alpha_i}} \cong \bigoplus_{j=1}^{l} C_{p^{\beta_j}}, \quad \alpha_1 \geqslant \alpha_2 \geqslant \cdots \geqslant \alpha_k \geqslant 1, \quad \beta_1 \geqslant \beta_2 \geqslant \cdots \geqslant \beta_l \geqslant 1,$$

where we may assume $\alpha_1 = \beta_1$. Now if $\alpha_1 = \beta_1 = 1$, we may regard T as a vector space over the field $D/(p)$, so that $k = l$.

[†]Here we use *linearly independent* in the broader sense that $\sum r_i a_i = 0$ implies that each $r_i a_i = 0$.

Now suppose that $\alpha_1 = \beta_1 \geqslant 2$ and that the theorem is proved for p-modules of p-exponent $< \alpha_1$. Moreover, assume that $\alpha_i \geqslant 2$ if and only if $i \leqslant m, \beta_j \geqslant 2$ if and only if $j \leqslant n$. Then

$$pT \cong \bigoplus_{i=1}^{m} C_{p^{\alpha_i - 1}} \cong \bigoplus_{j=1}^{n} C_{p^{\beta_j - 1}},$$

so that, by the inductive hypothesis, $m = n$, and $\alpha_i = \beta_i, 1 \leqslant i \leqslant m$.

Finally, $T/pT \cong \sum_{i=1}^{k} C_p \cong \sum_{j=1}^{l} C_p$. But T/pT is a vector space over $D/(p)$, so that $k = l$, and this completes the proof.

EXERCISES

1.1. Verify that $\mathbf{Z}_{(P)}$ and Γ are Euclidean domains. Show more generally that $\mathbf{Z}_{(P)}$ (Exercise 4.2 of Chapter 4) is a Euclidean domain.

1.2. Prove Proposition 1.3.

1.3. Let R be an integral domain and let S be a nonempty subset of R. We define the lcm (*least common multiple*) of S in R to be an element $m \in R$, if such exists, such that
(a) $a|m$ for all a in S.
(b) If $a|r$ for all a in $S, r \in R$, then $m|r$.
Show that the lcm, if it exists, is unique up to multiplication by a unit.

1.4. Show that the lcm always exists in a pid.

1.5. Let R be a pid, $a, b \in R$. Show that we may choose $d = \gcd(a, b), m = \mathrm{lcm}(a, b)$, so that

$$ab = dm.$$

1.6. Show that, in a pid, if $a|bc$ and $\gcd(a, b) = 1$, then $a|c$.

1.7. Show how, by an easy induction argument, we may prove the *existence* of a decomposition of a nonzero element of a Euclidean domain R without appealing to Lemma 1.7.

1.8. Find the gcd of $16 + 7i, 3 + 28i$ in the ring Γ of Gaussian integers. [*Hint*: apply the Euclidean function d, and use the fact that it satisfies (1.1).]

2.1. Show that there are ufd's in which Theorem 1.5 is false.

2.2. Find the gcd of $26x^2 - 104x + 104, 195x^2 + 65x - 910$ in $\mathbf{Z}[x]$.

2.3. Let R be a pid embedded in an integral domain D, let S be a family of elements of R and let $d = \gcd S$ in R. Show that $d = \gcd S$ in D.

2.4. Show that, in a ufd, if $a|bc$ and $\gcd(a, b) = 1$, then $a|c$.

2.5. Show that, in a ufd, every collection of elements has an lcm. Describe a process for computing the lcm if the collection is finite

Under what circumstances is the lcm of the collection S the zero element?

2.6. Show that, in a ufd, we may choose $d = \gcd(a,b), m = \operatorname{lcm}(a,b)$ so that $ab = dm$.

2.7. Express the quotient field construction as a left-adjoint functor to a certain underlying functor.

2.8. Generalize the construction of the quotient field as follows. Given an integral domain R, let S be a nonempty subset of R which is closed under multiplication and does not contain 0. Construct the set V of pairs (r,s) with $r \in R, s \in S$, and introduce an equivalence relation into V, and laws of addition and multiplication into the set $R[S^{-1}]$ of equivalence classes, just as in the construction of the quotient field. Show that $R[S^{-1}]$ is an integral domain, and that R may be embedded in $R[S^{-1}]$. Show that S has the required properties if it is the complement of a prime ideal I in R. We then say that $R[S^{-1}]$ is obtained from R by *localizing* at I. What do we get if we localize $\mathbf{Z}$ at the prime p?

3.1. Show that if R is Noetherian, then a direct sum of injective modules is injective. (Use Exercise 6.6 of Chapter 4.) (Note that this generalizes Exercise 6.2 of Chapter 4.)

3.2. Show that the converse of the statement in Exercise 3.1 is true. (Assume R to be non-Noetherian.)

3.3. If $R \to S$ is a ring homomorphism, show that
   (a) Every Noetherian S-module is R-Noetherian if and only if S is R-Noetherian.
   (b) $A \otimes_R S$ is a Noetherian S-module if A is a Noetherian R-module, provided that S is R-Noetherian.

3.4. An R-module A is called *artinian* if it satisfies the descending chain condition (DCC), that is, every chain

$$A_1 \supset A_1 \supset A_2 \supset \cdots$$

of submodules must stabilize after a finite number of terms. Show that
   (a) A is artinian if, and only if, every nonempty collection of submodules of A has a minimal element (see Theorem 3.1);
   (b) if $0 \to A' \to A \to A'' \to 0$ is an exact sequence of R-modules, then A is artinian $\Leftrightarrow A'$ and A'' are artinian (see Theorem 3.2).

3.5. For a ring R, let $J = J(R) = \cap$ {all maximal left ideals} be its *Jacobson radical*. Show that if $t \in J$, then $1 - t$ has a left inverse.

3.6. Show that if a ring R is left artinian, that is, left artinian as R-module, then

(a) $J(R) = \cap$ {a finite number of maximal left ideals};

(b) $J(R)$ is *nilpotent*, that is, there is an integer n such that $J^n = 0$.

4.1. We described in Section 3 of Chapter 2 a process for embedding an abelian group in a divisible abelian group. Show that this process generalizes to embedding a module over an integral domain in a divisible module.

4.2. Extend the notion of *torsion element, torsion submodule* to modules over arbitrary integral domains. Do they extend further to modules over arbitrary commutative rings?

4.3. Show that, with the extended definitions adopted in Exercise 4.2, Proposition 4.3 remains valid.

4.4. Let A be the abelian group $A = \Pi_p \mathbf{Z}_p$, p prime. Show that the torsion subgroup of A is not a direct summand.

[*Hint*: Show that (i) the torsion subgroup T is $\oplus_p \mathbf{Z}_p$ and (ii) there can be no retraction of A onto T.]

4.5. Let D be a Euclidean domain, let F be a free finitely generated D-module, and let R be a submodule of D. Show that R is free and that we may find bases $(x_1, x_2, \ldots, x_m)$ for F, $(y_1, y_2, \ldots, y_n)$ for R, $m \geqslant n$, such that

$$y_i = d_i x_i, \qquad i = 1, 2, \ldots, n,$$

and

$$d_1 | d_2 | \ldots | d_n.$$

[*Hint*: adapt a standard argument (for example, in [8]) for the case $D = \mathbf{Z}$.]

4.6. Translate the conclusion of Exercise 4.5 into the language of matrices.

4.7. Let V be a finite dimensional vector space over the field K and let $T : V \rightarrow V$ be a linear transformation from V to V. Let $K[x]$ operate on V by setting $xv = Tv$. Show that V is a torsion $K[x]$-module. Now take $K = \mathbf{C}$, the field of complex numbers, and use Theorems 4.7 and 4.9 to obtain the *eigenvalues* and *eigenspaces* of T, and the *characteristic polynomial* and *minimal polynomial* of T.

6

SEMISIMPLE RINGS

In this chapter we study an important class of rings, using methods developed in Chapter 4. Section 1 is concerned with an important technical theorem.

1. THE MORITA THEOREM

Let R be a ring and A a left R-module. Now $\mathrm{End}_R A$ has a ring structure induced by composition of functions. Since we write functions on the *left* of their arguments, the natural definition of a product in $\mathrm{End}_R A$ is by the rule

$$(\alpha\beta)(a) = \alpha(\beta(a)), \quad \alpha, \beta \in \mathrm{End}_R A, \quad a \in A. \tag{1.1}$$

Let us call the resulting ring E. Now, we have often suggested that there are great advantages in writing functions on the *right* of their arguments, and nowhere would this be more true than in this section! In order to mitigate, so far as possible, the consequences of our "left-handed" notational convention, we introduce the ring S, which is merely the opposite ring of E,

$$S = E^0, \quad E = S^0. \tag{1.2}$$

Thus a left S-module is nothing but a right E-module, and a right S-module is nothing but a left E-module.

We are going to show that, under certain circumstances, the categories $\mathfrak{M}_R$ and $\mathfrak{M}_S$ are equivalent. To this end, we note that A is, in fact, an (R, E)-left module, since

$$rs(a) = s(ra), \quad r \in R, \quad s \in E.$$

Thus A is an (R, S)-bimodule. It follows that the dual module A^*

181

$= \mathrm{Hom}_R(A, R)$ is an (S, R)-bimodule. It follows that $A^* \otimes_R A$ is an (S, S)-bimodule and $A \otimes_S A^*$ is an (R, R)-bimodule.

In an earlier chapter [see (5.3) of Chapter 4] we introduced the map $\theta = \theta_{AA}$,

$$\theta : A^* \otimes_R A \to \mathrm{End}_R A, \tag{1.3}$$

given by $\theta(\phi \otimes b)a = \phi(a)b$, $a, b \in A$, $\phi \in A^*$.

We quote from Theorem 5.14 of Chapter 4 and the remark preceding it.

THEOREM 1.1. *The map* $\theta : A^* \otimes_R A \to S$ *is a morphism of* (S, S)-*bimodules, and is an isomorphism if* A *is finitely-generated projective.*

Actually our proof showed that θ is an isomorphism if and only if A is finitely generated projective. We now introduce a companion morphism to θ, namely,

$$\tau : A \otimes_S A^* \to R,$$

given by

$$\tau(a \otimes \phi) = \phi(a). \tag{1.4}$$

To see that τ is well-defined, we should be explicit about the defining relation for the tensor product *over* S. Remembering that $S = E^0$, this reads

$$s(a) \otimes \phi = a \otimes \phi s, \quad a \in A, \quad \phi \in A^*, \quad s \in \mathrm{End}_R A. \tag{1.5}$$

It is thus clear that τ is well-defined by (1.4). Moreover, τ is a map of (R, R)-bimodules, since

$$\tau(r_1(a \otimes \phi)r_2) = \tau(r_1 a \otimes \phi r_2) = (\phi r_2)(r_1 a) = \phi(r_1 a)r_2 = r_1 \phi(a)r_2.$$

Thus $\tau(A \otimes_S A^*)$ is an ideal in R, called the *trace ideal* of A, which we write $\mathrm{Tr}\, A$. We say that A is an *R-generator* if τ is an isomorphism. It turns out that this is equivalent to asking that $\mathrm{Tr}\, A = R$.

PROPOSITION 1.2. *If* $\tau : A \otimes_S A^* \to R$ *is surjective, it is injective.*

PROOF. Since τ is surjective, there is an element $\sum a_j \otimes \phi_j$ in $A \otimes_S A^*$ such that $\tau(\sum a_j \otimes \phi_j) = 1$. Now, suppose that $\tau(\sum b_i \otimes \psi_i) = 0$. We want to prove that $\sum b_i \otimes \psi_i = 0$, given, then, that $\sum \psi_i(b_i) = 0$.

Now $\sum \phi_j(a_j) = 1$, so that

$$\sum_i b_i \otimes \psi_i = \sum_{i,j} \phi_j(a_j)b_i \otimes \psi_i = \sum_{i,j} \theta(\phi_j \otimes b_i)a_j \otimes \psi_i$$

$$= \sum_{i,j} a_j \otimes \psi_i \theta(\phi_j \otimes b_i), \quad \text{by (1.5)}.$$

Thus, to prove that $\sum b_i \otimes \psi_i = 0$, it suffices to show that $\sum \psi_i \theta(\phi \otimes b_i) = 0$ for all ϕ. Now, if $a \in A$,

$$\left(\sum \psi_i \theta(\phi \otimes b_i) \right)(a) = \sum \psi_i(\phi(a)b_i) = \phi(a) \sum \psi_i(b_i) = 0.$$

REMARK. If we study the proof of Theorem 5.14 of Chapter 4 we also note that θ is an isomorphism if it is surjective. A similar argument to that of Proposition 1.2 would have been available, but, of course, our goal was a different one on Section 5 of Chapter 4.

We call A an *R-progenerator* if it is a finitely generated projective R-generator. Thus A is an R-progenerator if θ and τ are isomorphisms. We then easily prove

THEOREM 1.3. *Let* A *be an* R-*progenerator. Then the categories* $\mathfrak{M}_R$, $\mathfrak{M}_S$ *are equivalent under the functors*

$$FM = M \otimes_R A, \quad M \in \mathfrak{M}_R$$

$$GN = N \otimes_S A^*, \quad N \in \mathfrak{M}_S.$$

PROOF.

$$GFM = (M \otimes_R A) \otimes_S A^*$$

$$\cong M \otimes_R (A \otimes_S A^*)$$

$$\cong M \otimes_R R, \quad \text{since } A \text{ is an } R\text{-generator,}$$

$$\cong M,$$

and

$$FGN = (N \otimes_S A^*) \otimes_R A$$

$$\cong N \otimes_S (A^* \otimes_R A)$$

$$\cong N \otimes_S S, \quad \text{since } A \text{ is finitely generated projective,}$$

$$\cong N.$$

Now we recall that, by definition, $S = (\text{End}_R A)^0$, where $\text{End}_R A$ has the natural (left-handed) ring structure (1.1). Thus the next result is to be expected in just the form in which it is stated.

COROLLARY 1.4 $R \cong End_S A$, *under the map* $r \mapsto \xi_r$, *where* $\xi_r(a) = ra$.

PROOF. Since F is an equivalence of categories and an additive functor, we have, for any $M_1, M_2 \in \mathfrak{M}_R$,

$$F_* : \operatorname{Hom}_R(M_1, M_2) \cong \operatorname{Hom}_S(FM_1, FM_2), \qquad (1.6)$$

as abelian groups. In particular,

$$F_* : \operatorname{Hom}_R(R, R) \cong \operatorname{Hom}_S(A, A).$$

Since $R \cong \operatorname{Hom}_R(R, R)$, we obtain the isomorphism of abelian groups

$$R \cong \operatorname{Hom}_S(A, A).$$

Tracing through this isomorphism, we find that it associates with $r \in R$, the endomorphism $a \mapsto ra$ of A. This is obviously a map of rings when $\operatorname{Hom}_S(A, A)$ is given the ring structure analogous to (1.1).

Indeed, we see that we finally have perfect symmetry between R and S. Utilizing the evident natural isomorphism between A and A^{**}, we find that A^* is an S-progenerator, the roles of τ and θ being interchanged.

In the next section we will see how progenerators arise.

2. SEMISIMPLE RINGS

DEFINITION 2.1. Let R be a ring. A left R-module A is *simple* if it is nonzero and has no proper submodules, and *semisimple* if it is a direct sum of simple modules. The ring R is *left semisimple* if it is semisimple as left R-module.

We first prove a theorem characterizing semisimple modules.

THEOREM 2.1.   *The following statements about* $A \in {}_R\mathfrak{M}$ *are equivalent*:
  (*i*) A *is semisimple.*
  (*ii*) *Every submodule of* A *is a direct summand in* A.
  (*iii*) A *is a sum of simple submodules.*

Our real interest is in the equivalence of (i) and (ii); statement (iii) arises in the course of the proof of the equivalence of (i) and (ii).

PROOF.
  (i)⇒(ii) Let $A = \oplus_{i \in I} A_i$, where A_i is simple, and let B be a submodule of A. Consider subsets J of I such that $B \cap \oplus_{j \in J} A_j = (0)$. This collection of subsets of I is ordered by inclusion and the ordering is evidently inductive. Thus there exists a maximal subset K of I such that

$$B \cap \bigoplus_{k \in K} A_k = (0).$$

We claim that $A = B \oplus C$, where $C = \oplus_k A_k$. Of course, it is sufficient to prove that $A = B + C$. Now, for each $i \notin K$ consider the projection $p_i : B \cap (C \oplus A_i) \to A_i$. The image is nonzero; for $B \cap C = (0)$, but $B \cap (C \oplus A_i) \neq (0)$. Thus, A_i being simple, the image is the whole of A_i.

Now let $a \in A$. Then $a = x + y, x \in C, y \in \oplus_{i \notin K} A_i$. Let $y = y_1 + \cdots + y_n$, $y_s \in A_{i_s}, i_s \notin K$. For each s, we may, by the reasoning above, find $z_s \in B$ with $z_s = c_s + y_s, c_s \in C$. Thus

$$a = x + \sum_{s=1}^{n} y_s$$

$$= \sum_{s=1}^{n} z_s + \left(x - \sum_{s=1}^{n} c_s \right),$$

whence $a \in B + C$.

(ii)$\Rightarrow$(iii) We first observe that property (ii) is inherited by submodules of A. For if A_0 is a submodule of A and B_0 is a submodule of A_0, then B_0 is a submodule of A, hence, by (ii), $A = B_0 \oplus C$. But then, obviously, $A_0 = B_0 \oplus C_0$, where $C_0 = C \cap A_0$.

Now let A_0 be a nonzero submodule of A and let $a \neq 0, a \in A_0$. We consider the submodules of A_0 that do not contain a. There is evidently a maximal one, say B_0. By the argument above, $A_0 = B_0 \oplus C_0$; we claim that C_0 is simple. For, if not, let $C_0 = C_1 \oplus C_2$, neither C_i being zero (this is legitimate, since every submodule of C_0 is complemented). Then $A_0 = B_0 \oplus C_1 \oplus C_2$. By the maximality of B_0, $B_0 \oplus C_1$ and $B_0 \oplus C_2$ both contain a. This however quickly leads to the false conclusion $a \in B_0$.

Finally, let $\mathfrak{F}$ be the collection of submodules of A that are sums of simple modules. Then $\mathfrak{F}$ is ordered by inclusion and the ordering is inductive. Let A' be a maximal element in $\mathfrak{F}$; if $A' \neq A$, then $A = A' \oplus A''$, with $A'' \neq (0)$. As shown above, $A'' = B'' \oplus C''$, with C'' simple. Then $A' \oplus C'' \in \mathfrak{F}$, contradicting the maximality of A'. Thus $A' = A$.

(iii)$\Rightarrow$(i) Suppose that $A = \sum_{i \in I} A_i$, where A_i is simple. Let us consider subsets J of I such that $\sum_{j \in J} A_j = \oplus_{j \in J} A_j$. This collection of subsets of I is ordered by inclusion and, again, the ordering is evidently inductive. Thus there exists a maximal subset K of I such that, for example,

$$\sum_{k \in K} A_k = \bigoplus_{k \in K} A_k = A'.$$

We will show that $A' = A$. For this it suffices, of course, to show that

$A_i \subseteq A'$ for every i. Now $A_i \cap A'$, being a submodule of A_i, is either 0 or A_i. But if $A_i \cap A' = 0$, then $i \notin K$ and $A' + A_i = A' \oplus A_i$, contradicting the maximality of K. Thus $A_i \cap A' = A_i$, so that $A_i \subseteq A'$, as desired.

COROLLARY 2.2. *Let* $A' \rightarrowtail A \twoheadrightarrow A''$ *be a short-exact sequence of R-modules. If* A *is semisimple, so are* A' *and* A'', *and the sequence splits.*

PROOF.   By condition (ii) of Theorem 2.1, A' is a direct summand in A, so the sequence splits. Moreover, we observed in the course of the proof of Theorem 2.1 that property (ii) is inherited by submodules. Thus A' is semisimple. Since the sequence splits, A'' is isomorphic to a submodule of A so that, by the same token, A'' is semisimple.

Our next theorem exhibits the enormous strength of the hypothesis that a ring is semisimple.

THEOREM 2.3.   *Let* R *be a ring. Then the following statements are equivalent*:

(*i*) R *is left semisimple.*
(*ii*) *Every left* R-*module is semisimple.*
(*iii*) *Every left* R-*module is projective.*
(*iv*) *Every left* R-*module is injective.*
(*v*) *Every short exact sequence of left* R-*modules splits.*

PROOF
(i)⇒(ii) Since R is left simisimple, so is every free left R-module, hence, by Corollary 2.2, every left R-module.
(ii)⇒(v) Corollary 2.2.
(v)⇒(iii), (iv) Obvious.
(iii) or (iv)⇒(i) Obvious, in view of condition (ii) of Theorem 2.1.

It is plain that every field provides an example of a semisimple ring. However, we point out in the next theorem that there are more interesting examples!

THEOREM 2.4.   (Maschke).   *Let* G *be a group of order* n, *and let* F *be a field whose characteristic does not divide* n. *Then* F[G] *is* (*left and right*) *semisimple.*

PROOF.   We will be content to demonstrate that $F[G]$ is left semisimple. Let N be a submodule (left ideal) of $F[G]$. There is then certainly an F-map $\pi : F[G] \rightarrow N$ such that $\pi x = x, x \in N$. Define $\phi : F[G] \rightarrow N$ by

$$\phi x = \frac{1}{n} \sum_{\sigma \in G} \sigma \pi (\sigma^{-1} x), \quad x \in F[G].$$

This definition is valid since $1/n \in F$. Plainly $\phi x = x$ if $x \in N$. Thus it remains to show that ϕ is a module map, since we will then have demonstrated that N is a direct summand in $F[G]$. But ϕ is obviously an F-map, so that it remains only to show that ϕ is a G-map. Now, if $\tau \in G$,

$$\phi(\tau x) = \frac{1}{n} \sum_{\sigma \in G} \sigma \pi (\sigma^{-1} \tau x)$$

$$= \tau \left(\frac{1}{n} \sum_{\sigma \in G} \tau^{-1} \sigma \pi \left((\tau^{-1} \sigma)^{-1} x \right) \right)$$

$$= \tau \left(\frac{1}{n} \sum_{\sigma' \in G} \sigma' \pi (\sigma'^{-1} x) \right) = \tau \phi(x),$$

since, as σ describes G, so does $\sigma' = \tau^{-1} \sigma$.

We are going to make a study of semisimple and simple rings. To this end we first prove a lemma.

LEMMA 2.5. *Let* I, J *be minimal left ideals in the ring* R. *Then either* IJ = 0 *or* I $\cong$ J.

PROOF. Plainly IJ is a left ideal contained in J. Thus $IJ = 0$ or $IJ = J$. If $IJ = J$, there must exist $a \in J$ such that $Ia \neq 0$. But then Ia, being a left ideal contained in J, must be the whole of J. Thus the map $x \mapsto xa$ is a module map of I onto J; its kernel is an ideal in I; hence it is zero. The given map is thus an isomorphism between I and J.

PROPOSITION 2.6. *Let* R *be a left semisimple ring, so that* $R = \oplus_\alpha J_\alpha$, *where each* J_α *is a minimal left ideal. If* I *is any minimal left ideal, then* I *is isomorphic to some* J_α.

PROOF. It cannot be true that $IJ_\alpha = 0$ for all α, for then $IR = 0$, so that $I = 0$. Thus $IJ_\alpha \neq 0$ for some α, so that $I \cong J_\alpha$, by Lemma 2.5.

Now let us take the expression $R = \oplus_\alpha J_\alpha$ and collect together the ideals J_α into isomorphism classes. For a given β, let R_β be the direct sum of the ideal J_α isomorphic to J_β. Thus, as left R-modules,

$$R = \bigoplus_{\beta \in B} R_\beta. \tag{2.1}$$

THEOREM 2.7. *If* R *is left semisimple, then*
(i) *Each* R_β *in* (2.1) *is a two-sided ideal in* R.
(ii) *The set* B *is finite.*
(iii) $R_\beta R_\gamma = 0$ *if* $\beta \neq \gamma$.

PROOF. Assertion (iii) follows immediately from Lemma 2.5. Then $R_\beta R = R_\beta R_\beta \subseteq R_\beta$, so that $R_\beta R = R_\beta$ and R_β is a right ideal. Also $1 = e_1 + \cdots e_k$, $e_i \in R_{\beta_i}$, from (2.1). But then $x = xe_1 + \cdots + xe_k$, so that $\beta = (\beta_1, \beta_2, \ldots, \beta_k)$.

Rephrasing our conclusions, we may enunciate

THEOREM 2.8. *A left semisimple ring* R *is a finite direct product of rings* R_β *which are left semisimple and have all their minimal left ideals isomorphic.*

PROOF. We have only to remark that the final phrase is justified by Proposition 2.6.

This theorem suggests the way we should define a left simple ring.

DEFINITION 2.2. A left semisimple ring is *left simple* if all its minimal left ideals are isomorphic.

PROPOSITION 2.9. *Let* R *be a left simple ring and* I *a nonzero left ideal. Then* $IR = R$.

PROOF. We may evidently suppose I to be minimal. Now $R = \oplus_i J_i$, where the J_i are mutually isomorphic minimal left ideals. We will show that, for each i, there exists $r_i \in R$ with $Ir_i = J_i$—this clearly suffices.

Let $\pi : R \to I$ be a projection and let $\phi : I \to J_i$ be an isomorphism, which exists by Proposition 2.6. Set $r_i = \phi\pi(1)$. Then $\phi\pi(x) = xr_i$, so that $\phi y = yr_i$, $y \in I$. Thus $J_i = Ir_i$, as claimed.

THEOREM 2.10. *A left simple ring contains no proper ideals.*

PROOF. Were I a nonzero (two-sided) ideal, then $I = IR = R$.

THEOREM 2.11. *A left simple module over a left simple ring is a progenerator.*

PROOF. Let R be left simple and let A be a left simple module over R. Then A is projective by Theorem 2.3, and it is plainly finitely generated because it is, in fact, cyclic.

It remains to show that the trace ideal of A is the whole of R. By Theorem 2.10 it thus suffices to show that $\text{Tr} A \neq (0)$. However, this is obvious, since A has a projective basis (a, f), $a \in A$, $f \in A^*$, and $f(a) \neq 0$.

Notice that a left simple module over R is isomorphic to a minimal left ideal of R.

THEOREM 2.12. (Wedderburn). *Let* R *be a left simple ring, and let* A *be a left simple* R*-module. Then, if* $S = (End_R A)^0$, $R \cong End_S A$, *where the isomorphism is defined by associating with* r *the left multiplication map* $a \mapsto ra$.

PROOF. We just apply Corollary 1.4.

PROPOSITION 2.13. (Schur). *The ring* $S = (End_R A)^0$ *is a division ring.*

PROOF. Since A is simple, $\text{End}_R A$ consists of zero and isomorphisms. Now any module over a division ring is free—the argument is just as for modules over fields, that is, vector spaces. Moreover, it is clear that A is a finitely generated S-module (since A^* certainly is), so that Theorem 2.12 and Proposition 2.13 may be combined in the following more classical form.

THEOREM 2.14. *Let* R *be a left simple ring. Then* R *is isomorphic to the ring of* $(n \times n)$-*matrices (for some* n) *over a division ring* S.

COROLLARY 2.15. *A ring is left semisimple if and only if it is right semisimple.*

PROOF. The converse of Theorem 2.14 is very straightforward. However, it is plain that we may replace "left" by "right" in that theorem. Thus a ring is left simple if and only if it is right simple. However, the converse of Theorem 2.8 is also immediate: a finite direct product of left simple rings is left semisimple. This proves the corollary.

(We have here an interesting contrast with Noetherian rings, studied in Chapter 5; there are rings that are left Noetherian without being right Noetherian.)

EXERCISES

1.1. Verify that the trace ideal $\tau(A \otimes_S A^*)$ is indeed an ideal of R.

1.2. Let $A \in {}_R\mathfrak{M}$. Show that the following are equivalent.
  (a) A is a generator.
  (b) R is a direct summand of a direct sum of a finite number of copies of A.
  (c) If $f: B \to C$ is nonzero, then there is an $\alpha: A \to B$ such that $f\circ\alpha \neq 0$.
  (d) $\text{Hom}_R(A, -)$ reflects isomorphisms, that is, if $f_*: \text{Hom}_R(A, B) \cong \text{Hom}_R(A, C)$, then $f: B \cong C$.
  (e) $\text{Hom}_R(A, -)$ reflects epimorphisms.

1.3. If A is an R-generator, show that every R-module is a quotient of a direct sum of copies of A.

1.4. Let $A \in {}_R\mathfrak{M}, S = \text{Hom}_R(A, A)$ and $C = \text{Hom}_S(A, A)$. Show that
  (a) $S = \text{Hom}_C(A, A)$.
  (b) A finitely generated projective R-module is an S-generator.
  (c) An R-generator is finitely generated in $\mathfrak{M}_S$.
   Give examples to show that the converse of (b) and (c) are false. [Take $R = \mathbf{Z}, A = \mathbf{Z}_2$.]

1.5. If A is an R-generator and if $\theta: R \to S$ is a ring homomorphism, show that $A \otimes_R S$ is an S-generator.

1.6. Let A be an R-module. Let $\text{Ann}_R A = \{r|rA = 0\}$. Show that $\text{Ann}_R A = \text{Ker}(\rho : R \to \text{End}_Z(A))$ where ρ is the module structure map. Thus $\text{Ann}_R A$ is an ideal, called the *annihilator ideal* of A. A is called *faithful* if $\text{Ann}_R A = 0$. Let R be a commutative ring and A a finitely generated R-module. Show that, for an ideal $I \subset R$ to have the property that $IA = A$, it is necessary and sufficient that $R = I + \text{Ann}_R(A)$. [Let $x_1, x_2, \ldots, x_n$ generate A and let $A_i = Rx_i + \cdots + Rx_n$, $A_{n+1} = 0$. Show, by induction, that there exists $a_i \in I$ such that $(1 - a_i)A \subset A_i$.]

1.7. (a) If A is a finitely generated projective R-module, show that $\text{Tr}(A)A = A$.

   (b) If, in addition, R is commutative, show that A is a generator if and only if A is faithful (see Exercise 1.6 for definition).

1.8. (Nakayama) Let A be a finitely generated R-module. Let $I \subset J(R)$ be an ideal of R. Show that, if $IA = A$, then $A = 0$. (Use Exercise 3.5 of Chapter 5.)

2.1. Show that a semisimple module is artinian if and only if it is Noetherian.

2.2. Show that an R-module is simple if and only if it is a quotient of R by a maximal left ideal.

2.3. Let R be a left artinian ring, with Jacobson radical J.

   (a) Show that R/J is semisimple. (Use Exercise 3.6 of Chapter 5 to produce an exact sequence

$$0 \to J \to R \to \bigoplus_{i=1}^{n} R/m_i$$

   where m_i are maximal left ideals.)

   (b) If $J^n = 0$ (Exercise 3.6 of Chapter 5), show that J^{i-1}/J^i, $i = 2, \ldots, n$, are artinian and semisimple. (J^{i-1}/J^i can be made into an A/J module.)

   (c) Show that R is left Noetherian.

2.4. (An alternative proof of Maschke's theorem) Let R be a semisimple ring and G a group. Then the ring homomorphism $R \to R[G]$ gives rise to a pair of adjoint functors $F : {}_R\mathfrak{M} \to {}_G\mathfrak{M}$ and $U : {}_G\mathfrak{M} \to {}_R\mathfrak{M}$. [$F = R[G] \otimes_R -$.] Show that

   (a) F carries projectives to projectives.

   (b) $R[G]$ is semisimple if and only if G is finite and G is a unit in R. [If $R[G]$ is semisimple, then $R[G] \to R$ splits, by t(say). Let $t(1) = \sum r_\sigma \sigma$. Show that, since $t(\sigma 1) = t(1)$, G must be finite and $r_\sigma = 1/|G|$. Conversely, define t as above, and set $s = t \otimes_R A : A \to R[G] \otimes_R A$.]

7

THE FUNCTORS
Ext AND Tor

1. CHAIN COMPLEXES, CHAIN MAPS, AND HOMOLOGY

A *chain complex* $\mathbf{C}$ of R-modules is a collection of R-modules C_n, $n \in \mathbf{Z}$, and module maps $\partial_n : C_n \to C_{n-1}$, $n \in \mathbf{Z}$, such that, for all n,

$$\partial_{n-1} \partial_n = 0. \tag{1.1}$$

The homomorphisms ∂_n are called *differentials* or *boundary operators*.

Let $Z_n = \ker \partial_n$, $B_n = \operatorname{im} \partial_{n+1}$. We call Z_n the module of *n-cycles*, B_n the module of *n-boundaries* of $\mathbf{C}$. Plainly, (1.1) implies that $B_n \subseteq Z_n$. The quotient module

$$H_n(\mathbf{C}) = Z_n / B_n \tag{1.2}$$

is called the nth *homology* module of C; two cycles determining the same element of H_n are called *homologous* and we denote by $[z]$ the homology class of the cycle z.

Plainly the homology module $H_n(C)$ measures the failure of $\mathbf{C}$ to be exact at C_n; precisely, $H_n(\mathbf{C}) = (0)$ if and only if $\mathbf{C}$ is exact at C_n.

To construct a category of chain complexes, we must define chain maps. A *chain map* $\phi : \mathbf{C} \to \mathbf{D}$ is a collection of module maps $\phi_n : C_n \to D_n$ such that, for each n, the diagram

$$
\begin{array}{ccc}
C_n & \xrightarrow{\phi_n} & D_n \\
\downarrow{\scriptstyle \partial_n} & & \downarrow{\scriptstyle \partial_n} \\
C_{n-1} & \xrightarrow{\phi_{n-1}} & D_{n-1}
\end{array}
\tag{1.3}
$$

commutes. Notice that we denote the differentials in $\mathbf{C}$ and $\mathbf{D}$ by the generic symbol ∂. Indeed, we often write conditions (1.3) simply as $\partial\phi = \phi\partial$ and say that "ϕ commutes with the differentials." (All the terms here are borrowed from algebraic topology.) Clearly, we may add chain maps $\mathbf{C} \to \mathbf{D}$, so that the collection of chain maps $\mathbf{C} \to \mathbf{D}$ forms an abelian group.

It is plain what we understand by the category of $\mathbf{Z}$-*graded* R-modules, $_R\mathfrak{M}^{\mathbf{Z}}$. We then have

PROPOSITION 1.1. *Chain complexes and chain maps constitute (under the usual law of function composition) an additive category* $_R\mathfrak{C}\mathfrak{h}$; *and homology is an additive functor*

$$H : {}_R\mathfrak{C}\mathfrak{h} \to {}_R\mathfrak{M}^{\mathbf{Z}}$$

PROOF. If $\phi : \mathbf{C} \to \mathbf{D}$ is a chain map, then clearly $\phi_n Z_n(\mathbf{C}) \subseteq Z_n(\mathbf{D})$, $\phi_n B_n(\mathbf{C}) \subseteq B_n(\mathbf{D})$, so that ϕ induces $H_n(\phi) : H_n(\mathbf{C}) \to H_n(\mathbf{D})$. The association $\phi \mapsto H_n(\phi)$ is obviously functorial. We will often write ϕ_* for $H(\phi)$, perhaps with a suitable subscript n.

Given two chain complexes $\mathbf{C}, \mathbf{C}'$, we may form a new chain complex $\mathbf{C}''$ by the rule $C_n'' = C_n \oplus C_n'$, $\partial_n'' = \partial_n \oplus \partial_n'$. It is plain that $\mathbf{C}''$ is the coproduct (direct sum) of $\mathbf{C}$ and $\mathbf{C}'$, $\mathbf{C}'' = \mathbf{C} \oplus \mathbf{C}'$ and that

$$H(\mathbf{C} \oplus \mathbf{C}') = H(\mathbf{C}) \oplus H(\mathbf{C}'). \tag{1.4}$$

We will also be concerned with *cochain complexes*. The difference between cochain complexes and chain complexes is simply this: whereas in the latter the differential lowers the *dimension* or *degree*, n, by 1, in the former it raises the degree by 1. We signal this difference by using the symbol δ for the differential in a cochain complex $\mathbf{C}$ and—often, but not always—by writing the degree symbol as a superscript instead of a subscript:

$$\delta^n : C^n \to C^{n+1}.$$

All the terms used for chain complexes ("cycles," "boundaries," "homology," and so on) are carried over to cochain complexes, each being preceded by the prefix "co." Plainly the difference between chain complexes and cochain complexes is purely formal, so that we do not need a separate theory of cochain complexes. Thus we will expect the reader to be able to make the automatic transition from statements about chain complexes to statements about cochain complexes. However, there is an important case in which the transition must be made with special care. We will be much concerned with *positive chain complexes*, that is, chain complexes **C** such that $C_n = 0, n < 0$. The theory of positive cochain complexes is then not identical with that of positive chain complexes but, rather, dual to it.

We close this section with some examples of chain and cochain complexes.

EXAMPLES.

1. Every module may be regarded as a chain, or cochain complex, *concentrated in dimension* 0. Thus, given the module A, we define $C_0 = A$, $C_n = (0)$, $n \neq 0$, $\partial_n = 0$—and similarly for the cochain complex. This device embeds $_R\mathfrak{M}$ in $_R\mathfrak{C}h$ as a full subcategory.

2. Let A be a module and let $F \twoheadrightarrow A$ be a free (or projective) presentation of A with kernel R, so that $R \overset{\mu}{\rightarrowtail} F \twoheadrightarrow A$ is a short exact sequence. Form the chain complex **C**, such that $C_1 = R$, $C_0 = F$, $C_n = (0)$, $n \neq 0, 1$, and $\partial_1 = \mu$. Then

$$H_0(\mathbf{C}) = A, \quad H_n(\mathbf{C}) = 0, \quad n \neq 0.$$

3. Dually, let $A \rightarrowtail I \overset{\epsilon}{\twoheadrightarrow} J$ be an injective presentation of A, and form the cochain complex **C**, with $C^0 = I, C^1 = J, C^n = (0), n \neq 0, \delta^0 = \epsilon$. Then

$$H^0(\mathbf{C}) = A, \quad H^n(\mathbf{C}) = 0, \quad n \neq 0.$$

4. Let **C** be a chain complex of left R-modules and let A be a left R-module. We define a chain complex $\mathbf{D} = \mathrm{Hom}_R(A, \mathbf{C})$ of abelian groups by $D_n = \mathrm{Hom}_R(A, C_n)$, $\partial_n^D = \partial_{n*}^C : \mathrm{Hom}_R(A, C_n) \rightarrow \mathrm{Hom}_R(A, C_{n-1})$. Now let B be a left R-module. We define a cochain complex $\mathbf{E} = \mathrm{Hom}_R(\mathbf{C}, B)$ of abelian groups by $E^n = \mathrm{Hom}_R(C_n, B)$, $\delta^n = \partial_{n+1}^* : \mathrm{Hom}_R(C_n, B) \rightarrow \mathrm{Hom}_R(C_{n+1}, B)$.

5. Let **C** be a chain complex of left R-modules and let A be a right R-module. We define a chain complex $\mathbf{B} = A \otimes_R \mathbf{C}$ by $B_n = A \otimes_R C_n, \partial_n^B = 1_A \otimes \partial_n^A$.

6. We already saw how we could form the direct sum of two chain complexes. It is obvious that we can form the direct sum, and direct product, of any family of chain complexes, or of cochain complexes.

2. CHAIN HOMOTOPY AND THE FUNDAMENTAL LEMMAS

The homology functor of Proposition 1.1 is central to this chapter. It is then natural that we should ask when two chain maps $\phi, \psi : \mathbf{C} \rightarrow \mathbf{D}$ induce the same homology homomorphism. In order that this should happen it is plainly necessary and sufficient that, for each $z \in Z(C)$, there exist $d \in \mathbf{D}$ with

$$\phi z - \psi z = \partial d. \tag{2.1}$$

We now seek a condition on ϕ and ψ (or, rather, on $\phi - \psi$) that guarantees this and behaves well with respect to additive functors; it should, of course, also be a condition that is often realized in practice when $\phi_* = \psi_*$. Again our definition is inspired by algebraic topology.

DEFINITION 2.1.   Let $\phi, \psi : \mathbf{C} \to \mathbf{D}$. A *chain homotopy* $\chi : \mathbf{C} \to \mathbf{D}$ *from* ϕ *to* ψ is a family of module maps $\chi_n : C_n \to D_{n+1}$, such that

$$\phi_n - \psi_n = \partial_{n+1}\chi_n + \chi_{n-1}\partial_n \tag{2.2}$$

We usually write (2.2) as

$$\phi - \psi = \partial\chi + \chi\partial.$$

We may write $\chi : \phi \approx \psi$ or $\phi \overset{\chi}{\approx} \psi$, or, simply $\phi \approx \psi$ if we do not need to specify the homotopy χ, and we say that ϕ is *homotopic* to ψ.

PROPOSITION 2.2.   *The relation $\phi \approx \psi$ is an equivalence relation on the morphisms of $_R\mathfrak{Ch}(\mathbf{C},\mathbf{D})$. Moreover, if $\phi \approx \psi : \mathbf{C} \to \mathbf{D}$ and $\alpha \approx \beta : \mathbf{D} \to \mathbf{E}$, then $\alpha\phi \approx \beta\psi$.*

PROOF.   Plainly $0 : \phi \approx \phi$; if $\chi : \phi \approx \psi$, then $-\chi : \psi \approx \phi$; and if $\chi : \phi \approx \psi$, $\lambda : \psi \approx \theta$, then $\chi + \lambda : \phi \approx \theta$.

Also if $\chi : \phi \approx \psi$ and $\alpha : \mathbf{D} \to \mathbf{E}$, then $\alpha\chi : \alpha\phi \approx \alpha\psi$. For

$$\alpha\phi - \alpha\psi = \alpha(\partial\chi + \chi\partial) = \partial\alpha\chi + \alpha\chi\partial.$$

Similarly, if $\lambda : \alpha \approx \beta$ and $\psi : \mathbf{C} \to \mathbf{D}$, then $\lambda\psi : \alpha\psi \approx \beta\psi$. Thus

$$\alpha\chi + \lambda\psi : \alpha\phi \approx \beta\psi.$$

PROPOSITION 2.3.   *If $\phi \approx \psi : \mathbf{C} \to \mathbf{D}$, then $\phi_* = \psi_* : \mathrm{H}(\mathbf{C}) \to \mathrm{H}(\mathbf{D})$.*

PROOF.   If $\chi : \phi \approx \psi$, and $z \in \mathbf{Z}(\mathbf{C})$, $\phi z - \psi z = \partial\chi z$, since $\partial z = 0$.

Now suppose that we are given an *additive* functor $F: {}_R\mathfrak{M} \to {}_S\mathfrak{M}$. It is then plain that F extends to a functor, which we will also call F,

$$F: {}_R\mathfrak{Ch} \to {}_S\mathfrak{Ch} \qquad (2.3)$$

which is also additive. Examples 4 and 5 at the end of Section 1 were, in fact, examples of such functors F. (We can plainly allow F to be contravariant.)

PROPOSITION 2.4. *Let* $\phi \approx \psi : C \to D$ *and let* $F: {}_R\mathfrak{M} \to {}_S\mathfrak{M}$ *be additive. Then* $F\phi \approx F\psi$.

PROOF. If F is covariant, we have, from (2.2),

$$F\phi - F\psi = F(\phi - \psi) = F(\partial\chi + \chi\partial) = (F\partial)(F\chi) + (F\chi)(F\partial).$$

But $F(C)$ is precisely the chain complex whose nth component is $F(C_n)$ and whose nth differential is $F\partial_n$. Thus we infer that if $\chi : \phi \approx \psi$, then $F\chi : F\phi \approx F\psi$. A slight modification of the argument is needed if F is contravariant.

It is important to note (see Exercise 2.1) that $\phi_* = \psi_*$ does *not* imply that $(F\phi)_* = (F\psi)_*$.

Proposition 2.2 may be paraphrased by saying that we have a category of chain complexes and homotopy classes of chain maps, ${}_R\mathfrak{Ch}_h$, called the *homotopy category*, and a *homotopy classification functor*

$$Q: {}_R\mathfrak{Ch} \to {}_R\mathfrak{Ch}_h.$$

Then Proposition 2.3 asserts that H factors (uniquely) through $Q, H = \overline{H}Q$.

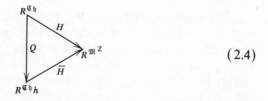

$$(2.4)$$

We may write $[\phi]$ for the homotopy class of ϕ. Equivalent objects of ${}_R\mathfrak{Ch}$ are of the same *homotopy type*. An object equivalent to O is *contractible*.

We note that we may go one stage further and construct *homotopies between homotopies*. This idea (which also owes its origin to topological homotopy theory) will be exploited in Theorem 2.6.

We now prove two fundamental lemmas of homological algebra. We will also give their dual forms, but we will be content to prove them as first stated. We will be concerned with *positive* chain complexes. We say that such a complex **P** is *pseudoprojective* if each P_n is projective.

THEOREM 2.5. *Let* **C** *be a positive chain complex. Then there exists a pseudoprojective positive chain complex* **P** *and a chain map* $\phi : \mathbf{P} \to \mathbf{C}$ *inducing an isomorphism in homology.*

PROOF. Of course, we define $P_n = (0), \phi_n = 0$ if $n < 0$. Now suppose that we have constructed **P** and ϕ up to dimension $n - 1, n \geqslant 0$, so that

$$\phi_{n-1*} : H_{n-1}(\mathbf{P}) \twoheadrightarrow H_{n-1}(\mathbf{C}),$$

$$\phi_{r*} : H_r(\mathbf{P}) \cong H_r(\mathbf{C}), \quad r \leqslant n - 2.$$

Consider the diagram

$$
\begin{array}{ccc}
Q & \dashrightarrow[\theta] & C_n \\
\downarrow{\scriptstyle \epsilon} & \phi & \downarrow{\scriptstyle \partial} \\
K & \longrightarrow & B_{n-1}(\mathbf{C}) \\
\downarrow & & \downarrow \\
Z_{n-1}(\mathbf{P}) & \xrightarrow{\phi} & Z_{n-1}(\mathbf{C}) \\
\downarrow{\scriptstyle \zeta\phi} & & \downarrow{\scriptstyle \zeta} \\
H_{n-1}(\mathbf{C}) & = & H_{n-1}(\mathbf{C})
\end{array}
$$

Here ζ projects $Z_{n-1}(\mathbf{C})$ onto $H_{n-1}(\mathbf{C})$. Since ϕ_{n-1*} is surjective, it follows that $\zeta\phi$ is surjective. Let K be the kernel of $\zeta\phi$. Then $\phi : P_{n-1} \to C_{n-1}$ sends K into $B_{n-1}(\mathbf{C})$. Let $\epsilon : Q \twoheadrightarrow K$ be a projective presentation of K. Since ∂ maps C_n onto $B_{n-1}(\mathbf{C})$, we may find $\theta : Q \to C_n$ with $\partial\theta = \phi\epsilon$.

Now let $\eta : R \twoheadrightarrow Z_n(\mathbf{C})$ be a projective presentation of $Z_n(\mathbf{C})$. We set

$$P_n = Q \oplus R, \quad \partial_n = \langle \epsilon, 0 \rangle : P_n \to P_{n-1}, \quad \phi_n = \langle \theta, \eta \rangle : P_n \to C_n.$$

Then $\partial_n \phi_n = \phi_{n-1} \partial_n$, since $\partial\eta = 0$ and $\partial\theta = \phi\epsilon$. Also $\partial_{n-1}\partial_n = 0 : P_n \to P_{n-2}$ since $\epsilon Q \subseteq Z_{n-1}(\mathbf{P})$. Also, with this modified complex **P** we have $H_{n-1}(\mathbf{P}) = Z_{n-1}(\mathbf{P})/K$, so that

$$\phi_{n-1*} : H_{n-1}(\mathbf{P}) \cong H_{n-1}(\mathbf{C}).$$

Finally

$$\phi_{n*}: H_n(\mathbf{P}) \twoheadrightarrow H_n(\mathbf{C}),$$

since $R \subseteq Z_n(\mathbf{P})$ and $\phi_n R = Z_n(\mathbf{C})$. This completes the inductive step, and hence establishes the theorem.

Our next theorem shows that the choice of $\mathbf{P}$ in Theorem 2.5 is very heavily restricted.

THEOREM 2.6. *In the diagram of positive chain complexes and chain maps,*

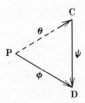

ψ *induces homology isomorphisms and* $\mathbf{P}$ *is pseudoprojective. Then there exists* θ, *unique up to homotopy, such that* $\psi\theta \approx \phi$.

PROOF. We first prove the existence of θ. We define $\theta_n = 0$, $\chi_n = 0 : P_n \to D_{n+1}$, $n < 0$, and we assume inductively that we have defined

$$\theta_r : P_r \to C_r, \quad \chi_r : P_r \to D_{r+1}, \quad r \leqslant n-1, \quad n \geqslant 0$$

such that

$$\partial\theta_r = \theta_{r-1}\partial, \quad \phi_r - \psi_r\theta_r = \partial\chi_r + \chi_{r-1}\partial, \quad r \leqslant n-1.$$

Now

$$\partial(\phi_n - \chi_{n-1}\partial) = (\phi_{n-1} - \partial\chi_{n-1})\partial = (\psi_{n-1}\theta_{n-1} + \chi_{n-2}\partial)\partial = \psi_{n-1}\theta_{n-1}\partial.$$

Thus, since

$$\partial\theta_{n-1}\partial = \theta_{n-2}\partial\partial = 0,$$

$$\theta_{n-1}\partial : P_n \to Z_{n-1}(\mathbf{C}) \quad \text{and} \quad \psi_{n-1}\theta_{n-1}\partial : P_n \to B_{n-1}(\mathbf{D}). \quad (2.5)$$

Since ψ induces an injection in homology, it follows from (2.5) that, in fact,

$$\theta_{n-1}\partial : P_n \to B_{n-1}(\mathbf{C}).$$

Since P_n is projective we may find $\bar{\theta} : P_n \to C_n$ with $\theta_{n-1}\partial = \partial\bar{\theta}$. Then

$$\partial(\phi_n - \chi_{n-1}\partial) = \psi_{n-1}\partial\bar{\theta} = \partial\psi_n\bar{\theta},$$

so that

$$\phi_n - \chi_{n-1}\partial - \psi_n\bar{\theta} : P_n \to Z_n(\mathbf{D}). \tag{2.6}$$

Since ψ induces a surjection in homology, it follows from (2.6), and the fact that P_n is projective, that there exist $\rho : P_n \to Z_n(\mathbf{C})$, $\chi_n : P_n \to D_{n+1}$, with

$$\phi_n - \chi_{n-1}\partial - \psi_n\bar{\theta} = \psi_n\rho + \partial\chi_n.$$

Set $\theta_n = \bar{\theta} + \rho$. Then $\partial\theta_n = \partial\bar{\theta} = \theta_{n-1}\partial$ and

$$\phi_n - \psi_n\theta_n = \partial\chi_n + \chi_{n-1}\partial.$$

This establishes the inductive step, hence the existence of θ. The proof of uniqueness is very similar. It is plain that all we have to prove is that if $\psi\theta \approx 0$, then $\theta \approx 0$. Thus we suppose that we have a chain homotopy $\sigma : \psi\theta \approx 0$, that is,

$$\psi\theta = \partial\sigma + \sigma\partial.$$

We will construct a homotopy $\tau : \theta \approx 0$ and a homotopy of homotopies $\gamma : \sigma \approx \psi\tau$, inductively. Precisely (starting as usual with zero in negative dimensions), we suppose that we have defined

$$\tau_r : P_r \to C_{r+1}, \quad \gamma_r : P_r \to D_{r+2}, \quad r \leqslant n-1, \quad n \geqslant 0,$$

so that

$$\theta_r = \partial\tau_r + \tau_{r-1}\partial, \quad \sigma_r - \psi_{r+1}\tau_r = \partial\gamma_r - \gamma_{r-1}\partial, \quad r \leqslant n-1.$$

Then

$$\partial(\theta_n - \tau_{n-1}\partial) = (\theta_{n-1} - \partial\tau_{n-1})\partial = 0,$$

so that

$$\theta_n - \tau_{n-1}\partial : P_n \to Z_n(\mathbf{C}) \tag{2.7}$$

and

$$\psi_n(\theta_n - \tau_{n-1}\partial) = \partial\sigma_n + \sigma_{n-1}\partial + (\partial\gamma_{n-1} - \gamma_{n-2}\partial - \sigma_{n-1})\partial = \partial(\sigma_n + \gamma_{n-1}\partial),$$

so that

$$\psi_n(\theta_n - \tau_{n-1}\partial) : P_n \to B_n(\mathbf{D}). \tag{2.8}$$

Since ψ induces an injection in homology, it follows from (2.7) and (2.8) that, in fact,

$$\theta_n - \tau_{n-1}\partial : P_n \to B_n(\mathbf{C}).$$

Since P_n is projective, we fine $\bar{\tau} : P_n \to C_{n+1}$ with $\theta_n - \tau_{n-1}\partial = \partial\bar{\tau}$. Then

$$\partial\psi_{n+1}\bar{\tau} = \psi_n\partial\bar{\tau} = \psi_n(\theta_n - \tau_{n-1}\partial) = \partial(\sigma_n + \gamma_{n-1}\partial),$$

so that

$$\sigma_n - \psi_{n+1}\bar{\tau} + \gamma_{n-1}\partial : P_n \to Z_{n+1}(\mathbf{D}). \tag{2.9}$$

Since ψ induces a surjection in homology, it follows from (2.9), and the fact that P_n is projective, that there exist $\kappa : P_n \to Z_{n+1}(\mathbf{C})$, $\gamma_n : P_n \to D_{n+2}$, with

$$\sigma_n - \psi_{n+1}\bar{\tau} + \gamma_{n-1}\partial = \psi_{n+1}\kappa + \partial\gamma_n.$$

Set $\tau_n = \bar{\tau} + \kappa$. Then $\theta_n = \tau_{n-1}\partial + \partial\bar{\tau} = \tau_{n-1}\partial + \partial\tau_n$, and

$$\sigma_n - \psi_{n+1}\tau_n = \partial\gamma_n - \gamma_{n-1}\partial.$$

This establishes the inductive step, and hence completes the proof of the theorem.

COROLLARY 2.7. *Let* $\mathbf{P}, \mathbf{Q}$ *be pseudoprojective positive complexes and let* $\phi : \mathbf{P} \to \mathbf{Q}$ *induce homology isomorphisms. Then* ϕ *is a homotopy equivalence (that is, equivalence in* $_R\mathfrak{Ch}_h$*).*

PROOF. By Theorem 2.6, there exists $\psi : \mathbf{Q} \to \mathbf{P}$ (unique up to homotopy) with $\phi\psi \approx 1 : \mathbf{Q} \to \mathbf{Q}$. But then $\phi\psi\phi \approx \phi$. By the uniqueness part of Theorem 2.6, we infer that $\psi\phi \approx 1$.

COROLLARY 2.8. *Let* $\mathbf{P}, \mathbf{Q}$ *be pseudoprojective positive complexes, let* $\mathbf{C}$ *be a positive complex, and let* $\phi : \mathbf{P} \to \mathbf{C}, \psi : \mathbf{Q} \to \mathbf{C}$ *induce homology isomorphisms. Then there exists a homotopy equivalence* $\tau : \mathbf{P} \approx \mathbf{Q}$*, unique up to homotopy, such that* $\psi\tau \approx \phi$*.*

PROOF. The existence and uniqueness of τ follow directly from Theorem 2.6. Since $\psi_*\tau_* = \phi_*$ and ψ_*, ϕ_* are isomorphisms, so is τ_*. Thus τ is a homotopy equivalence by Corollary 2.7.

COROLLARY 2.9. *Let* $\mathbf{C}$ *be a positive chain complex with* $H(\mathbf{C}) = (0)$*.*

Then, for all pseudoprojective positive complexes **P** *and all chain maps* $\phi : \mathbf{P} \rightarrow \mathbf{C}$ *we have* $\phi \approx 0$.

PROOF. $0 : \mathbf{C} \rightarrow \mathbf{O}$ induces homology isomorphisms. We apply the uniqueness part of Theorem 2.6.

COROLLARY 2.10. *Let* **P** *be a pseudoprojective positive chain complex. Then* **P** *is contractible if and only if* $H(\mathbf{P}) = 0$.

PROOF. If $1 \approx 0 : \mathbf{P} \rightarrow \mathbf{P}$ then certainly $1 = 0 : H(\mathbf{P}) \rightarrow H(\mathbf{P})$, so $H(\mathbf{P}) = 0$. Conversely, if $H(\mathbf{P}) = 0$, then $1 \approx 0 : \mathbf{P} \rightarrow \mathbf{P}$ by Corollary 2.9.

THEOREM 2.11. *Let* $_R\mathfrak{Ch}_h^+$ *be the homotopy category of positive chain complexes and let* $_R\mathfrak{P}_h^+$ *be the full subcategory of pseudoprojective complexes. Then the embedding functor* $E : {}_R\mathfrak{P}_h^+ \subseteq {}_R\mathfrak{Ch}_h^+$ *admits a right adjoint* G, *such that* $\overline{H}G = \overline{H}$, *and* $G^2 = G$.

PROOF. This very abstract formulation conceals what is, in the light of Theorems 2.5 and 2.6, a simple fact. Given **C**, an object of $_R\mathfrak{Ch}_h^+$, we may associate with it the complex **P** of Theorem 2.5. By Corollary 2.8, **P** is determined up to canonical equivalence in $_R\mathfrak{P}_h^+$. This determines the functor G. Obviously $\overline{H}G = \overline{H}$ and $G^2 = G$. The adjointness simply expresses the fact that $\phi : \mathbf{P} \rightarrow \mathbf{C}$ in Theorem 2.5 sets up a one-one correspondence between homotopy classes of chain maps $\mathbf{Q} \rightarrow \mathbf{C}$ and homotopy classes of chain maps $\mathbf{Q} \rightarrow \mathbf{P}$, provided that **Q** is pseudoprojective; this fact is guaranteed by Theorem 2.6.

The "dual" theory is concerned with positive *cochain* complexes **C** and *pseudoinjective* complexes **I**, which are positive cochain complexes such that I^n is injective for every *n*. We merely state the duals of Theorems 2.5 and 2.6 and leave the proofs to the reader. In the Exercises, we will also ask the reader to state and prove the duals of the consequences we have explicitly drawn from Theorems 2.5 and 2.6.

THEOREM 2.5*. *Let* **C** *be a positive cochain complex. Then there exists a pseudoinjective positive cochain complex* **I** *and a cochain map* $\phi : \mathbf{C} \rightarrow \mathbf{I}$ *inducing an isomorphism in cohomology.*

THEOREM 2.6*. *In the diagram of positive cochain complexes and cochain maps*

ψ induces cohomology isomorphisms and **I** *is pseudoinjective. Then there exists* θ, *unique up to homotopy, such that* θψ ≈ φ.

3. THE FUNCTOR Ext

Here we specialize the considerations of Section 2. Let A be an R-module; we may regard A as a chain complex concentrated in dimension 0 (see Section 1, Example 1). Then we use Theorem 2.5 to find a pseudoprojective complex **P** and a chain map $\phi : \mathbf{P} \to A$ inducing homology isomorphisms. This means that

$$\cdots \to P_{n+1} \xrightarrow{\partial_{n+1}} P_n \xrightarrow{\partial_n} P_{n-1} \to \cdots \to P_0 \xrightarrow{\phi_0} A \to 0$$

is exact, or that

$$H_n(\mathbf{P}) = 0, \quad n > 0; \quad \phi_* : H_0(\mathbf{P}) \cong A. \tag{3.1}$$

We call such a complex **P** a *projective resolution* of A; more strictly, it is the pair $(\mathbf{P}, \phi)$ which constitutes the resolution. Of course, it is much easier to prove the existence of a projective resolution than to establish the more general Theorem 2.5.

Now, by Corollary 2.8, if $(\mathbf{P}, \phi)$ and $(\mathbf{P}', \phi')$ are any two projective resolutions of A, there exists a homotopy equivalence τ, unique up to homotopy,

$$\tau : \mathbf{P} \approx \mathbf{P}', \tag{3.2}$$

such that $\phi' \tau \approx \phi$. However, it is plain that chain maps into A are homotopic if and only if they are equal, so that $\phi' \tau = \phi$.

Consider now the cochain complex $\operatorname{Hom}_R(\mathbf{P}, B)$, for a given R-module B (see Section 1, Example 4). Since $\operatorname{Hom}_R(-, B)$ is an additive functor, it follows (from Proposition 2.4) that

$$\operatorname{Hom}_R(\tau, B) : \operatorname{Hom}_R(\mathbf{P}', B) \approx \operatorname{Hom}_R(\mathbf{P}, B). \tag{3.3}$$

Thus $\operatorname{Hom}_R(\tau, B)$ induces an isomorphism in cohomology. Since τ is determined up to homotopy by the two resolutions, this means that there is a *canonical* isomorphism between $H^*(\operatorname{Hom}_R(\mathbf{P}', B))$ and $H^*(\operatorname{Hom}_R(\mathbf{P}, B))$, by which we may identify them. Thus we may define abelian groups, $\operatorname{Ext}_R^n(A, B)$, which depend only on A and B as follows.

DEFINITION 3.1.

$$\text{Ext}_R^n (A,B) = H^n(\text{Hom}_R (\mathbf{P},B)),$$

where $\phi: \mathbf{P} \to A$ is an arbitrary projective resolution of A.
Actually we may go further and prove

THEOREM 3.1. *$Ext_R^n(-,-)$ is an additive bifunctor, contravariant in the first variable, and covariant in the second variable.*

PROOF. We must show how to associate with $\alpha: A_2 \to A_1$ a homomorphism

$$\alpha^*: \text{Ext}_R^n (A_1,B) \to \text{Ext}_R^n (A_2,B) \tag{3.4}$$

and with $\beta: B_1 \to B_2$ a homomorphism

$$\beta_*: \text{Ext}_R^n (A,B_1) \to \text{Ext}_R^n (A,B_2) \tag{3.5}$$

in such a way that (with the evident meanings)

$$1^* = 1, \quad (\alpha\alpha')^* = \alpha'^*\alpha^*, \tag{3.6}$$

$$1_* = 1, \quad (\beta'\beta)_* = \beta'_*\beta_*. \tag{3.7}$$

$$\alpha^*\beta_* = \beta_*\alpha^*. \tag{3.8}$$

We must then also show that

$$\text{Ext}_R^n (A_1 \oplus A_2,B) = \text{Ext}_R^n (A_1,B) \oplus \text{Ext}_R^n (A_2,B) \tag{3.9}$$

$$\text{Ext}_R^n (A,B_1 \oplus B_2) = \text{Ext}_R^n (A,B_1) \oplus \text{Ext}_R^n (A,B_2). \tag{3.10}$$

The definition of β_* and the functorial property (3.7) are dealt with in Exercise 1.2, and are, in any case, quite obvious—likewise (3.10). Also (3.9) presents no problem; for if $\phi_{(i)}: \mathbf{P}_{(i)} \to A_i$ are projective resolutions, $i = 1, 2$, then $\phi_{(1)} \oplus \phi_{(2)}: \mathbf{P}_{(1)} \oplus \mathbf{P}_{(2)} \to A_1 \oplus A_2$ is plainly a projective resolution, and

$$\text{Hom}_R (\mathbf{P}_{(1)} \oplus \mathbf{P}_{(2)},B) = \text{Hom}_R (\mathbf{P}_{(1)},B) \oplus \text{Hom}_R (\mathbf{P}_{(2)},B). \tag{3.11}$$

Thus (3.9) follows from (3.11) by passing to cohomology.
We now deal with the crucial matter—the definition of α^*. Our reasoning is again based on Theorem 2.6. Let $\phi_{(i)}: \mathbf{P}_{(i)} \to A_i$ be projective resolu-

tions, $i = 1, 2$, and let $\alpha : A_2 \rightarrow A_1$. By Theorem 2.6, there exists

$$\psi : \mathbf{P}_{(2)} \rightarrow \mathbf{P}_{(1)},$$

unique up to homotopy, such that

$$\phi_{(1)}\psi = \alpha\phi_{(2)}. \tag{3.12}$$

Note that we can write equality in (3.12), since, as already remarked, there are no nontrivial homotopies into a module (chain complex concentrated in dimension 0). We use ψ to define α^*. Thus

$$\mathrm{Hom}_R (\psi, B) : \mathrm{Hom}_R (\mathbf{P}_{(1)}, B) \rightarrow \mathrm{Hom}_R (\mathbf{P}_{(2)}, B),$$

and, since ψ is determined up to homotopy by α, the cohomology homomorphism induced by $\mathrm{Hom}_R(\psi, B)$ is determined by α. We call this cohomology homomorphism α^*.

Strictly speaking, α^* depends on the choices of resolution, but we have already seen how to pass from one resolution to another in canonical fashion. The functorial property is now obvious; for if we have also $\alpha' : A_3 \rightarrow A_2$ and a resolution $\phi_{(3)} : \mathbf{P}_{(3)} \rightarrow A_3$, then we find ψ', unique up to homotopy, such that

$$\phi_{(2)}\psi' = \alpha'\phi_{(3)}.$$

Then

$$\phi_{(1)}\psi\psi' = \alpha\alpha'\phi_{(3)},$$

so that, by definition, $(\alpha\alpha')^*$ is the cohomology homomorphism induced by

$$\mathrm{Hom}_R (\psi\psi', B) = \mathrm{Hom}_R(\psi', B)\mathrm{Hom}_R (\psi, B). \tag{3.13}$$

Since passage to (co) homology is functorial, (3.13) implies that $(\alpha\alpha')^*$ $= \alpha'^*\alpha^*$. Of course, the relation $1^* = 1$ is trivially true.

It remains to establish (3.8). But this springs from the evident commutativity of the square

$$
\begin{array}{ccc}
\mathrm{Hom}_R(\mathbf{P}_{(1)}, B_1) & \xrightarrow{\mathrm{Hom}_R(\mathbf{P}_{(1)}, \beta)} & \mathrm{Hom}_R(\mathbf{P}_{(1)}, B_2) \\
\Big\downarrow {\scriptstyle \mathrm{Hom}_R(\psi, B_1)} & & \Big\downarrow {\scriptstyle \mathrm{Hom}_R(\psi, B_2)} \\
\mathrm{Hom}_R(\mathbf{P}_{(2)}, B_1) & \xrightarrow[\mathrm{Hom}_R(\mathbf{P}_{(2)}, \beta)]{} & \mathrm{Hom}_R(\mathbf{P}_{(2)}, B_2)
\end{array}
\tag{3.14}
$$

In other words, (3.8) is a simple reflection of the bifunctoriality of $\text{Hom}_R(-,-)$.

Actually, Hom_R is a special case of Ext_R^n, as we now show.

THEOREM 3.2.

$$\text{Ext}_R^0(A,B) \cong \text{Hom}_R(A,B).$$

PROOF. Let $\phi : \mathbf{P} \to A$ be a projective resolution of A. If we look at the bottom dimensions of $\mathbf{P}$, we see that this implies that $P_1 \xrightarrow{\partial} P_0 \xrightarrow{\phi_0} A \to 0$ is right exact. Since $\text{Hom}_R(-,B)$ is a left-exact functor, it follows that

$$0 \to \text{Hom}_R(A,B) \to \text{Hom}_R(P_0,B) \xrightarrow{\delta^0} \text{Hom}_R(P_1,B)$$

is left exact. Thus $\text{Hom}_R(A,B) \cong \ker \delta^0 = H^0(\text{Hom}_R(P,B)) = \text{Ext}_R^0(A,B)$.

Now, explicitly,

$$\text{Ext}_R^n(A,B) = \ker \delta^n / \text{im}\, \delta^{n-1}, \quad n \geqslant 1, \tag{3.15}$$

in the sequence

$$\text{Hom}_R(P_{n-1},B) \xrightarrow{\delta^{n-1}} \text{Hom}_R(P_n,B) \xrightarrow{\delta^n} \text{Hom}_R(P_{n+1},B);$$

here, of course, $\delta^i = \text{Hom}_R(\partial_{i+1},B)$, $\partial_{i+1} : P_{i+1} \to P_i$, $i = n-1, n$. We use (3.15) to prove

THEOREM 3.3. $\text{Ext}_R^n(A,B) \cong \text{Hom}_R(B_{n-1},B)/i^* \text{Hom}_R(P_{n-1},B)$, $n \geqslant 1$, where $B_{n-1} = B_{n-1}(\mathbf{P})$, and $i : B_{n-1} \subseteq P_{n-1}$.

PROOF. The sequence $P_{n+1} \xrightarrow{\partial_{n+1}} P_n \xrightarrow{\bar{\partial}} B_{n-1} \to 0$ is right exact, where we have factored ∂_n as

$$P_n \xrightarrow{\bar{\partial}} B_{n-1} \xrightarrow{i} P_{n-1}. \tag{3.17}$$

Thus $0 \to \text{Hom}_R(B_{n-1},B) \xrightarrow{\bar{\partial}^*} \text{Hom}_R(P_n,B) \xrightarrow{\delta^n} \text{Hom}_R(P_{n+1},B)$ is left exact, and so $\bar{\partial}^*$ maps $\text{Hom}_R(B_{n-1},B)$ isomorphically onto $\ker \delta^n$. Since $\delta^{n-1} = \partial_n^* = \bar{\partial}^* i^*$, it follows that $\bar{\partial}^*$ maps $\text{im}\, i^*$ isomorphically onto $\text{im}\, \delta^{n-1}$. This completes the proof.

We now show how to reduce $\text{Ext}_R^n(A,B)$, $n \geqslant 1$, to $\text{Ext}_R^1(X,B)$, for a suitable X.

PROPOSITION 3.4. *Let* $\phi : \mathbf{P} \to A$ *be a projective resolution of* A, *and let* $n \geqslant 1$. *Define* $\mathbf{Q}$ *by*

$$Q_i = P_{n+i}; \quad \partial_i^Q = \partial_{n+i}^P, \quad i \geqslant 1,$$

and define $\psi : \mathbf{Q} \to B_{n-1}$ *by* $\psi_0 = \bar{\partial} : P_n \twoheadrightarrow B_{n-1}$ (3.17). *Then* $(\mathbf{Q}, \psi)$ *is a projective resolution of* B_{n-1}.

The proof is immediate.

COROLLARY 3.5.

$$\operatorname{Ext}_R^{n+1}(A,B) \cong \operatorname{Ext}_R^1(B_{n-1}, B), \quad n \geqslant 1.$$

PROOF.   Note that $B_i(\mathbf{Q}) = B_{n+i}(\mathbf{P}), i \geqslant 0$. Thus, by Theorem 3.3,

$$\operatorname{Ext}_R^1(B_{n-1}, B) \cong \operatorname{Hom}_R(B_0(\mathbf{Q}), B)/i^* \operatorname{Hom}_R(Q_0, B)$$

$$= \operatorname{Hom}_R(B_n(\mathbf{P}), B)/i^* \operatorname{Hom}_R(P_n, B)$$

$$\cong \operatorname{Ext}_R^{n+1}(A,B), \quad \text{again by Theorem 3.3.}$$

Note that the same argument yields the more general result

$$\operatorname{Ext}_R^n(A,B) \cong \operatorname{Ext}_R^{n-k}(B_{k-1}, B), \quad n \geqslant 2, \quad i \leqslant k \leqslant n-1. \tag{3.6}$$

There is a dual development which we merely outline here. Regard the module B as a cochain complex concentrated in dimension 0. We apply Theorem 2.5[*] to obtain a cochain map $\phi : B \to \mathbf{I}$ inducing cohomology isomorphisms. This means that

$$\phi_{0*} : B \cong H^0(\mathbf{I}), \quad \text{and} \quad H^i(\mathbf{I}) = 0, \quad i \geqslant 1. \tag{3.18}$$

We call $\mathbf{I}$ or, more strictly, $\phi : B \to \mathbf{I}$, an *injective resolution* of B. We then make the following definition of $\overline{\operatorname{Ext}}_R^n(A,B)$.

DEFINITION 3.1[*]

$$\overline{\operatorname{Ext}}_R^n(A,B) = H^n(\operatorname{Hom}_R(A,\mathbf{I})),$$

where $\phi : B \to \mathbf{I}$ is an arbitrary injective resolution of B. We then may prove (and the reader should!)

THEOREM 3.1*. $\overline{\text{Ext}}\,_{R}^{n}(-,-)$ *is an additive bifunctor, contravariant in the first variable, and covariant in the second.*

Actually, it will turn out that $\text{Ext}_{R}^{n}(-,-)$ and $\overline{\text{Ext}}\,_{R}^{n}(-,-)$ are naturally equivalent. However, we require considerable preparation before we are in a position to establish this important result.

THEOREM 3.2*.

$$\overline{\text{Ext}}\,_{R}^{0}(A,B) \cong \text{Hom}_{R}(A,B).$$

THEOREM 3.3*.

$$\overline{\text{Ext}}\,_{R}^{n}(A,B) \cong \text{Hom}_{R}(A,B^{n-1})/j_{*}\text{Hom}_{R}(A,I^{n-1}), \quad n \geqslant 1,$$

where $B^{n-1} = im\,\delta^{n-1} : I^{n-1} \to I^{n}$, *and* j *projects* I^{n-1} *onto* B^{n-1}.

COROLLARY 3.5*.

$$\overline{\text{Ext}}\,_{R}^{n+1}(A,B) \cong \overline{\text{Ext}}\,_{R}^{1}(A,B^{n-1}), \quad n \geqslant 1.$$

The reader should prove these results, guided by the arguments in the dual case.

4. PROPERTIES OF Ext

Our objective in this section is to prove the existence of two basic exact sequences related to the Ext functor. The proofs are, however, based on more general theorems of homological algebra which we now enunciate and prove in the generality in which we will need them.

THEOREM 4.1. **(The Snake Lemma).** *Consider the commutative diagram in* $_{R}\mathfrak{M}$

$$
\begin{array}{ccccc}
K' & \overset{\mu_1}{\to} & K & \overset{\eta_1}{\to} & K'' \\
\downarrow{\kappa'} & & \downarrow{\kappa} & & \downarrow{\kappa''} \\
A' & \overset{\mu_2}{\to} & A & \overset{\eta_2}{\to} & A'' & \to 0 \\
\downarrow{f'} & & \downarrow{f} & & \downarrow{f''} \\
0 \to & B' & \overset{\mu_3}{\to} & B & \overset{\eta_3}{\to} & B'' \\
\downarrow{\rho'} & & \downarrow{\rho} & & \downarrow{\rho''} \\
C' & \overset{\mu_4}{\to} & C & \overset{\eta_4}{\to} & C''
\end{array}
$$

*where the two middle rows and the columns are exact. Then there is a
homomorphism* $\Delta : K'' \to C'$ *such that the sequence*

$$K' \xrightarrow{\mu_1} K \xrightarrow{\eta_1} K'' \xrightarrow{\Delta} C' \xrightarrow{\mu_4} C \xrightarrow{\eta_4} C'' \qquad (4.2)$$

is exact.

Note that the top row of (4.1) is just the "kernel row" of (f',f,f''); and
the bottom row is just the "cokernel row". Thus the snake lemma is about
a map of a right exact sequence into a left exact sequence.

PROOF.   We first prove exactness at K. Since $0 = \eta_2 \mu_2 \kappa' = \kappa'' \eta_1 \mu_1$ and κ''
is monic, it follows that $\eta_1 \mu_1 = 0$. Now let $\eta_1 x = 0, x \in K$. Then $\eta_2 \kappa x = \kappa'' \eta_1 x$
$= 0$, so that $\kappa x = \mu_2 a', a' \in A'$. But $\mu_3 f' a' = f \mu_2 a' = f \kappa x = 0$, so that, μ_3 being
monic, $f'a' = 0$, hence $a' = \kappa' y, y \in K'$. Thus $\kappa x = \mu_2 \kappa' y = \kappa \mu_1 y$ and, κ being
monic, $x = \mu_1 y$.

Next we define Δ. Let $x'' \in K''$. Then $\kappa'' x'' = \eta_2 a$, for some $a \in A$.
Moreover, $\eta_3 f a = f'' \eta_2 a = f'' \kappa'' x'' = 0$, so that $fa = \mu_3 b', b' \in B'$. We set $\Delta x''$
$= \rho' b'$. We exercised choice, however, in picking a so that $\eta_2 a = \kappa'' x''$.
Thus, to justify our definition of Δ, we must show that $\rho' b'$ is independent
of the choice of a. Now we may alter a to $a_1 = a + \mu_2 a'$, where a' is an
arbitrary element of A'. Then $fa_1 = fa + f\mu_2 a' = fa + \mu_3 f' a'$, so that b' is
altered to $b_1' = b' + f'a'$. But then $\rho' b_1' = \rho' b'$. It is plain that Δ, being
well-defined, is a homomorphism of R-modules; for, being composed of
homomorphisms and their inverses, it is a homomorphism, provided that it
is a function. (Indeed, we call this theorem the *Snake Lemma* because Δ is
obtained by zigzagging through (4.1), $\Delta = \rho' \mu_3^{-1} f \eta_2^{-1} \kappa''$.)

We next prove exactness at K''. If $x'' = \eta_1 x$, $x'' \in K''$, $x \in K$, we choose
κx as that element $a \in A$ such that $\eta_2 a = \kappa'' x''$. Then $fa = 0$, so $\Delta x'' = 0$.
Thus $\Delta \eta_1 = 0$. Conversely, let $\Delta x'' = 0$. Then $\kappa'' x'' = \eta_2 a$, $fa = \mu_3 b', \rho' b' = 0$.
Thus $b' = f'a', a' \in A'$, so that $fa = \mu_3 f' a' = f \mu_2 a'$. It follows that $\eta_2 a_1 = \kappa'' x''$,
$fa_1 = 0$, where $a_1 = a - \mu_2 a'$. Then $a_1 = \kappa x, \kappa'' \eta_1 x = \kappa'' x''$, so that $x'' = \eta_1 x$.
We leave to the reader the proofs of exactness of (4.2) at C' and C.

REMARKS

(i) We note that μ_1 is monic if μ_2 is monic, and η_4 is epic if η_3 is epic.

(ii) The *connecting homomorphism* Δ is plainly natural in the following
sense. We consider the part of (4.1) consisting of the two middle rows; call
this (f',f,f''). It is then plain what is meant by a map of (f',f,f'') into
(f_1',f_1,f_1''). Such a map induces maps $\beta : K'' \to K_1'', \gamma : C' \to C_1'$ (in obvious
notation) and Δ is natural in the sense that

$$
\begin{array}{ccc}
K'' & \xrightarrow{\Delta} & C' \\
{\scriptstyle \beta} \downarrow & & \downarrow {\scriptstyle \gamma} \\
K_1'' & \xrightarrow{\Delta_1} & C_1'
\end{array}
$$

commutes. The reader is recommended to give the detailed proof of this assertion.

From Theorem 4.1 we deduce

THEOREM 4.2.   *Let* $0 \to C' \overset{\phi}{\to} C \overset{\psi}{\to} C'' \to 0$ *be a short exact sequence of chain complexes. Then there exists a connecting homomorphism* $\omega_n : H_n(C'') \to H_{n-1}(C')$ *such that the sequence*

$$\cdots \to H_n(C') \overset{\phi_*}{\to} H_n(C) \overset{\psi_*}{\to} H_n(C'') \overset{\omega}{\to} H_{n-1}(C') \overset{\phi_*}{\to} \cdots$$

is exact.

PROOF.   Applying the Snake Lemma to

$$
\begin{array}{ccccccccc}
0 & \to & C'_r & \overset{\phi_r}{\to} & C_r & \overset{\psi_r}{\to} & C''_r & \to & 0 \\
 & & \downarrow{\scriptstyle\partial} & & \downarrow{\scriptstyle\partial} & & \downarrow{\scriptstyle\partial} & & \\
0 & \to & C'_{r-1} & \overset{\phi_{r-1}}{\to} & C_{r-1} & \overset{\psi_{r-1}}{\to} & C''_{r-1} & \to & 0
\end{array}
$$

we infer [see Remark (i) following Theorem 4.1] that the sequences

$$0 \to Z'_r \overset{\phi_r}{\to} Z_r \overset{\psi_r}{\to} Z''_r, \quad C'_{r-1}/B'_{r-1} \overset{\phi_{r-1}}{\to} C_{r-1}/B_{r-1} \overset{\psi_{r-1}}{\to} C''_{r-1}/B''_{r-1} \to 0,$$

are exact for all r. Furthermore, the boundary homomorphism induces homomorphisms, which we also write as ∂,

$$
\begin{array}{ccccccc}
C'_n/B'_n & \overset{\phi_n}{\to} & C_n/B_n & \overset{\psi_n}{\to} & C''_n/B''_n & \to & 0 \\
\downarrow{\scriptstyle\partial} & & \downarrow{\scriptstyle\partial} & & \downarrow{\scriptstyle\partial} & & \\
0 \to Z'_{n-1} & \overset{\phi_{n-1}}{\to} & Z_{n-1} & \overset{\psi_{n-1}}{\to} & Z''_{n-1} & &
\end{array}
\tag{4.4}
$$

Moreover, in (4.4) the kernel of $\partial : C_n/B_n \to Z_{n-1}$ is $H_n(C)$, and the cokernel is $H_{n-1}(C)$. Thus the theorem follows by applying the Snake Lemma to (4.4).

REMARK.   Just as in the Snake Lemma, we may infer that ω is natural in an evident sense.

We are now ready to prove the existence of the two basic exact sequences related to the Ext functor.

THEOREM 4.3.   *Let* $0 \to B' \overset{\mu}{\to} B \overset{\epsilon}{\to} B'' \to 0$ *be a short exact sequence of*

R-*modules. Then there is an exact sequence*

$$0 \to \operatorname{Hom}_R (A, B') \xrightarrow{\mu_*} \operatorname{Hom}_R (A, B) \xrightarrow{\epsilon_*} \operatorname{Hom}_R (A, B'') \xrightarrow{\omega} \operatorname{Ext}_R^1 (A, B') \to \cdots$$

$$\to \operatorname{Ext}_R^n (A, B') \xrightarrow{\mu_*} \operatorname{Ext}_R^n (A, B) \xrightarrow{\epsilon_*} \operatorname{Ext}_R^n (A, B'') \xrightarrow{\omega} \operatorname{Ext}_R^{n+1} (A, B') \to \cdots$$

PROOF. Let $\mathbf{P} \to A$ be a projective resolution of A. Since each P_n is projective, it follows that the sequence

$$0 \to \operatorname{Hom}_R (\mathbf{P}, B') \xrightarrow{\mu_*} \operatorname{Hom}_R (\mathbf{P}, B) \xrightarrow{\epsilon_*} \operatorname{Hom}_R (\mathbf{P}, B'') \to 0 \quad (4.5)$$

is also short exact. We may thus apply Theorem 4.2 to obtain the theorem (in view of Theorem 3.2).

THEOREM 4.4. *Let* $0 \to A' \xrightarrow{\mu} A \xrightarrow{\epsilon} A'' \to 0$ *be a short exact sequence of* R-*modules. Then there is an exact sequence.*

$$0 \to \operatorname{Hom}_R (A'', B) \xrightarrow{\epsilon^*} \operatorname{Hom}_R (A, B) \xrightarrow{\mu^*} \operatorname{Hom}_R (A', B) \xrightarrow{\omega} \operatorname{Ext}_R^1 (A'', B) \to \cdots$$

$$\operatorname{Ext}_R^n (A'', B) \xrightarrow{\epsilon^*} \operatorname{Ext}_R^n (A, B) \xrightarrow{\mu^*} \operatorname{Ext}_R^n (A', B) \xrightarrow{\omega} \operatorname{Ext}_R^{n+1} (A'', B) \to \cdots.$$

PROOF. Here the application of Theorem 4.2 is not quite so straight-forward. We need a preliminary lemma.

LEMMA 4.5. *Given* $0 \to A' \xrightarrow{\mu} A \xrightarrow{\epsilon} A'' \to 0$ *exact, we may construct a commutative diagram*

$$
\begin{array}{ccccccccc}
0 & \to & \mathbf{P}' & \xrightarrow{\mu} & \mathbf{P} & \xrightarrow{\epsilon} & \mathbf{P}'' & \to & 0 \\
& & \downarrow & & \downarrow & & \downarrow & & \\
0 & \to & A' & \xrightarrow{\mu} & A & \xrightarrow{\epsilon} & A'' & \to & 0
\end{array}
\quad (4.6)
$$

where each column is a projective resolution of the module at its base and the top row is exact.

PROOF. Take arbitrary projective presentations $P_0' \xrightarrow{\eta'} A'$, $P_0'' \xrightarrow{\eta''} A''$ and set $P_0 = P_0' \oplus P_0''$. Since P_0'' is projective, we may find $\theta : P_0'' \to A$ with $\epsilon\theta = \eta''$. Set $\eta = \langle \mu\eta', \theta \rangle : P_0 \to A$. We have a diagram

$$P_0' \overset{\mu_0}{\rightarrowtail} P_0 \overset{\epsilon_0}{\twoheadrightarrow} P_0'' \tag{4.7}$$

$$\downarrow \eta' \qquad \downarrow \eta \qquad \downarrow \eta''$$

$$A' \overset{\mu}{\rightarrowtail} A \overset{\epsilon}{\twoheadrightarrow} A''$$

where μ_0, ϵ_0 are the canonical injection and projection, and it is easy to see that (4.7) commutes. It now follows that η is surjective, since the induced sequence

$$\operatorname{coker} \eta' \to \operatorname{coker} \eta \to \operatorname{coker} \eta''$$

is exact by Theorem 4.1. Thus $\eta : P_0 \twoheadrightarrow A$ is a projective presentation. The Snake Lemma now permits us to infer from (4.7) an exact sequence

$$\ker \eta' \rightarrowtail \ker \eta \twoheadrightarrow \ker \eta'', \tag{4.8}$$

and we repeat the reasoning above with (4.8) replacing the original sequence $A' \rightarrowtail A \twoheadrightarrow A''$. Proceeding in this way, we build up the diagram (4.6).

We now return to the proof of Theorem 4.4. Since each P_n'' is projective, the sequences $0 \to P_n' \to P_n \to P_n'' \to 0$ split so that exactness is preserved when we apply the contravariant functor $\operatorname{Hom}_R(-, B)$. Thus we have a short exact sequence

$$0 \to \operatorname{Hom}_R(\mathbf{P}'', B) \overset{\epsilon^*}{\to} \operatorname{Hom}_R(\mathbf{P}, B) \overset{\mu^*}{\to} \operatorname{Hom}_R(\mathbf{P}', B) \to 0, \tag{4.9}$$

to which we may apply Theorem 4.2. This yields the exact sequence of the theorem.

We may also record the corresponding results for $\overline{\operatorname{Ext}}$; later, these will be superfluous, since we will have identified Ext and $\overline{\operatorname{Ext}}$.

THEOREM 4.3*. *Let* $0 \to A' \overset{\mu}{\to} A \overset{\epsilon}{\to} A'' \to 0$ *be a short exact sequence of R-modules. Then there is an exact sequence*

$$0 \to \operatorname{Hom}_R(A'', B) \overset{\epsilon^*}{\to} \operatorname{Hom}_R(A, B)$$

$$\overset{\mu^*}{\to} \operatorname{Hom}_R(A', B) \overset{\omega}{\to} \overline{\operatorname{Ext}}\,^1_R(A'', B) \to \cdots$$

$$\to \overline{\operatorname{Ext}}\,^n_R(A'', B) \overset{\epsilon^*}{\to} \overline{\operatorname{Ext}}\,^n_R(A, B) \overset{\mu^*}{\to} \overline{\operatorname{Ext}}\,^n_R(A', B) \overset{\omega}{\to} \overline{\operatorname{Ext}}\,_R^{n+1}(A'', B) \to \cdots.$$

THEOREM 4.4*. *Let* $0 \to B' \overset{\mu}{\to} B \overset{\epsilon}{\to} B'' \to 0$ *be a short exact sequence* *of* R-*modules. Then there is an exact sequence*

$$0 \to \mathrm{Hom}_R(A, B') \xrightarrow{\mu_*} \mathrm{Hom}_R(A, B) \xrightarrow{\epsilon_*} \mathrm{Hom}_R(A, B'') \xrightarrow{\omega} \overline{\mathrm{Ext}}\,{}^1_R(A, B') \to \cdots$$

$$\to \overline{\mathrm{Ext}}\,{}^n_R(A, B') \xrightarrow{\mu_*} \overline{\mathrm{Ext}}\,{}^n_R(A, B) \xrightarrow{\epsilon_*} \mathrm{Ext}^n_R(A, B'') \xrightarrow{\omega} \overline{\mathrm{Ext}}\,{}_R^{n+1}(A, B') \to \cdots.$$

We now draw some important consequences from our theorems.

THEOREM 4.6. *The following statements about the* R-*module* A *are equivalent*:

(*i*) A *is projective.*

(*ii*) $Ext^n_R(A, B) = 0$ *for all* B *and all* $n \geqslant 1$.

(*iii*) $Ext^1_R(A, B) = 0$ *for all* B.

PROOF.

(i)$\Rightarrow$(ii) If A is projective, we may choose a resolution P of A with $P_0 = A, P_n = 0, n \geqslant 1$. Thus (ii) follows.

(ii) $\Rightarrow$(iii) Trivial.

(iii)$\Rightarrow$(i) Let $B_0 \overset{i}{\rightarrowtail} P_0 \twoheadrightarrow A$ be a projective presentation of A. By Theorem 4.4, we have an exact sequence

$$0 \to \mathrm{Hom}_R(A, B) \to \mathrm{Hom}_R(P_0, B) \xrightarrow{i^*} \mathrm{Hom}_R(B_0, B) \to \mathrm{Ext}^1_R(A, B) \to \cdots.$$

But $\mathrm{Ext}^1_R(A, B) = 0$, so that i^* is surjective. Setting $B = B_0$, we find $\theta : P_0 \to B_0$ with $\theta i = 1$. Thus the sequence $B_0 \rightarrowtail P_0 \twoheadrightarrow A$ splits, so that A, as a direct summand in P_0, is projective.

THEOREM 4.7. *The following statements about the* R-*module* B *are equivalent*:

(*i*) B *is injective.*

(*ii*) $Ext^n_R(A, B) = 0$ *for all* A *and all* $n \geqslant 1$.

(*iii*) $Ext^1_R(A, B) = 0$ *for all* A.

PROOF.

(i)$\Rightarrow$(ii) We apply Theorem 3.3, observing that, since B is injective, every homomorphism from B_{n-1} to B extends to P_{n-1}.

(ii)$\Rightarrow$(iii) Trivial.

(iii)$\Rightarrow$(i) Let $B \rightarrowtail I^0 \xrightarrow{j} B^0$ be an injective presentation of B. By Theorem 4.3, we have an exact sequence

$$0 \to \mathrm{Hom}_R(A, B) \to \mathrm{Hom}_R(A, I^0) \xrightarrow{j_*} \mathrm{Hom}_R(A, B^0) \to \mathrm{Ext}^1_R(A, B) \to \cdots.$$

But $\mathrm{Ext}^1_R(A, B) = 0$, so that j_* is surjective. Setting $A = B^0$, we find

$\theta : B^0 \to I^0$ with $j\theta = 1$. Thus the sequence $B \rightarrowtail I^0 \twoheadrightarrow B^0$ splits, so that B, as a direct factor in I^0, is injective.

We will leave the duals of Theorems 4.6 and 4.7 implicit, but will feel free to apply them.

THEOREM 4.8.

$$\operatorname{Ext}^n_R (A, B) \cong \overline{\operatorname{Ext}}\,^n_R (A, B).$$

PROOF. If $n = 0$, we appeal to Theorems 3.2 and 3.2*. If $n = 1$, we apply Theorem 4.3* to the projective presentation $0 \to B_0 \xrightarrow{i} P_0 \to A \to 0$. We have an exact sequence

$$0 \to \operatorname{Hom}_R (A, B) \to \operatorname{Hom}_R (P_0, B) \xrightarrow{i^*} \operatorname{Hom}_R (B_0, B) \to \overline{\operatorname{Ext}}\,^1_R (A, B) \to 0$$

by Theorem 4.7*. Thus $\overline{\operatorname{Ext}}\,^1_R (A, B) \cong \operatorname{Hom}_R (B_0, B) / i^* \operatorname{Hom}_R (P_0, B)$ $\cong \operatorname{Ext}^1_R (A, B)$, by Theorem 3.3.

If $n \geqslant 2$, we proceed by induction. We have

$$\operatorname{Ext}^n_R (A, B) \cong \operatorname{Ext}^{n-1}_R (B_0, B), \quad \text{by } (3.6),$$

$$\cong \overline{\operatorname{Ext}}\,_R^{n-1} (B_0, B), \quad \text{by the inductive hypothesis,}$$

$$\cong \overline{\operatorname{Ext}}\,^n_R (A, B),$$

since $\ldots \to \overline{\operatorname{Ext}}\,^{n-1}_R (P_0, B) \to \overline{\operatorname{Ext}}\,^{n-1}_R (B_0, B) \to \overline{\operatorname{Ext}}\,^n_R (A, B) \to \overline{\operatorname{Ext}}\,^n_R (P_0, B) \to \cdots$ is exact (Theorem 4.3*) and $\overline{\operatorname{Ext}}\,^{n-1}_R (P_0, B)$ $= \overline{\operatorname{Ext}}\,^n_R (P_0, B) = 0$ (Theorem 4.7*).

REMARK. We have not proved an actual equivalence of bifunctors $\overline{\operatorname{Ext}}\,^n_R (-, -) \cong \overline{\operatorname{Ext}}\,^n_R (-, -)$. This involves a number of more or less automatic verifications which we will suppress. The diligent reader should convince himself on this point.

We next examine the dependence of $\operatorname{Ext}^n_R (A, B)$ on A and B. Precisely, we consider the following question: under what circumstances does a homomorphism $\phi : A \to A'$ induce an isomorphism $\phi^* : \operatorname{Ext}^n_R (A', B)$ $\cong \operatorname{Ext}^n_R (A, B)$ for all B and all $n \geqslant 1$. First, we assert that it is plain (see Corollary 3.5*) that this is equivalent to asking that

$$\phi^* : \operatorname{Ext}^1_R (A', B) \cong \operatorname{Ext}^1_R (A, B), \quad \text{for all } B.$$

We prove

THEOREM 4.9. *The homomorphism* $\phi : A \to A'$ *induces* $\phi^* : \operatorname{Ext}^1_R (A', B)$ $\cong \operatorname{Ext}^1_R (A, B)$ *for all* B *if and only if* ϕ *factors as*

$$A \overset{\phi_1}{\rightarrowtail} A \oplus P \overset{\phi_2}{\cong} A' \oplus P' \overset{\phi_3}{\twoheadrightarrow} A',$$

where P, P' *are projective,* ϕ_1 *is the canonical injection, and* ϕ_3 *is the canonical projection.*

PROOF. Since $\operatorname{Ext}^1_R (A \oplus P, B) = \operatorname{Ext}^1_R (A, B) \oplus \operatorname{Ext}^1_R (P, B) = \operatorname{Ext}^1_R (A, B)$ ((3.9) and Theorem 4.6), it is plain that ϕ_1 induces an isomorphism ϕ_1^*. Similarly ϕ induces an isomorphism ϕ_3^*. Since certainly ϕ induces an isomorphism ϕ_2^*, it follows that if $\phi = \phi \, \phi_2 \phi_1$, then $\phi^* = \phi_1^* \phi^{*2} \phi^{*3}$ is an isomorphism.

Conversely, suppose that $\phi^* : \operatorname{Ext}^1_R (A', B) \to \operatorname{Ext}^1_R (A, B)$ is an isomorphism for all B. We first suppose that ϕ is surjective with kernel A'',

$$A'' \overset{\psi}{\rightarrowtail} A \overset{\phi}{\twoheadrightarrow} A', \tag{4.10}$$

yielding

$$\cdots \to \operatorname{Hom}_R (A, B) \overset{\psi^*}{\to} \operatorname{Hom}_R (A'', B) \overset{\omega}{\to} \operatorname{Ext}^1_R (A', B) \overset{\phi^*}{\to} \operatorname{Ext}^1_R (A, B) \overset{\psi^*}{\to}$$

$$\operatorname{Ext}^1_R (A'', B) \overset{\omega}{\to} \operatorname{Ext}^2_R (A', B) \overset{\phi^*}{\to} \operatorname{Ext}^2_R (A, B). \tag{4.11}$$

Since ϕ^* is an isomorphism (both times!) in (4.11), it follows that $\operatorname{Ext}^1_R (A'', B) = 0$, for all B, so that A'' is projective by Theorem 4.6. It also follows that $\psi^* : \operatorname{Hom}_R (A, B) \to \operatorname{Hom}_R (A'', B)$ is surjective, so that, setting $B = A''$, we infer that (4.10) splits. This means that ϕ factors as $A \overset{\theta}{\cong} A' \oplus P' \overset{\phi_3}{\twoheadrightarrow} A'$, where $P' = A''$ and ϕ_3 is the canonical projection.

We now consider the general case $\phi : A \to A'$. Let $P \overset{\epsilon}{\twoheadrightarrow} A'$ be a projective presentation of A' and set $\bar{\phi} = \langle \phi, \epsilon \rangle : A \oplus P \to A'$. Then $\bar{\phi}$ is surjective. Moreover, $\phi = \bar{\phi} \phi_1$, where $\phi_1 : A \rightarrowtail A \oplus P$ is the canonical injection. We are given that ϕ^* is an isomorphism, and we have already observed that ϕ_1^* is an isomorphism; hence $\bar{\phi}^*$ is an isomorphism. Thus we may apply our previous reasoning to $\bar{\phi}$ to infer that $\bar{\phi}$ factors as

$$A \oplus P \overset{\phi_2}{\cong} A' \oplus P' \overset{\phi_3}{\twoheadrightarrow} A'.$$

Finally, then, $\phi = \bar{\phi} \phi_1$ factors as $\phi = \phi_3 \phi_2 \phi_1$ and the theorem is proved.

The dual form of Theorem 4.9 is

THEOREM 4.9*. *The homomorphism* $\psi : B \to B'$ *induces* $\psi_* : \operatorname{Ext}_R^1(A, B)$ $\cong \operatorname{Ext}_R^1(A, B')$ *for all* A *if and only if* ψ *factors as*

$$B \overset{\psi_1}{\rightarrowtail} B \oplus I \overset{\psi_2}{\cong} B' \oplus I' \overset{\psi_3}{\twoheadrightarrow} B',$$

where **I, I'** *are injective,* ψ_1 *is the canonical injection, and* ψ_3 *is the canonical projection.*

5. THE FUNCTOR Tor

The process implied by Definition 3.1 is also described as that of obtaining the *n*th *right derived functor* of the functor $\operatorname{Hom}_R(-, B)$. That is to say, we take the additive functor $F = \operatorname{Hom}_R(-, B)$ and we construct a new functor $R^n F$ whose value at A is given by

$$(R^n F)(A) = H^n(\operatorname{Hom}_R(\mathbf{P}, B)),$$

where **P** is a projective resolution of A. Similarly, Definition 3.1* indicates that $\overline{\operatorname{Ext}}_R^n(A, B)$ is the value at B of the *n*th right derived functor $R^n G$, where $G = \operatorname{Hom}_R(A, -)$. Then Theorem 4.7 asserts that

$$(R^n F)(A) = (R^n G)(B), \tag{5.1}$$

where $F = \operatorname{Hom}_R(-, B), G = \operatorname{Hom}_R(A, -)$.

We now describe a very similar development that may be made, based on the tensor product instead of Hom. Thus we now consider the two functors

$$F = A \otimes_R - : {}_R\mathfrak{M} \to \mathfrak{Ab}, \quad G = - \otimes_R B : \mathfrak{M}_R \to \mathfrak{Ab}, \tag{5.2}$$

where $A \in \mathfrak{M}_R$, $B \in {}_R\mathfrak{M}$. These are additive functors, and we will prove that their *n*th *left derived functors* stand in a relation analogous to (5.1). Thus we will define

$$\overline{\operatorname{Tor}}_n^R(A, B) = H_n(A \otimes_R \mathbf{Q}), \quad \operatorname{Tor}_n^R(A, B) = H_n(\mathbf{P} \otimes_R B), \tag{5.3}$$

where $\mathbf{P} \to A, \mathbf{Q} \to B$ are projective resolutions, and we will prove

$$\operatorname{Tor}_n^R(A, B) \cong \overline{\operatorname{Tor}}_n^R(A, B) \tag{5.4}$$

Indeed, it will again turn out that $\operatorname{Tor}_n^R(-, -)$ is a bifunctor; however, it is (unlike $\operatorname{Ext}_R^n(-, -)$) covariant in both variables.

In view of the extensive discussion of the Ext functor in the preceding two sections, it should not be necessary to go into great detail over Tor. Indeed, we follow an entirely parallel course to establish the facts. Thus, we first show that $\text{Tor}_n^R(A, B)$ is independent of the choice of projective resolution $\mathbf{P} \to A$. For, as in (3.2), given another $\mathbf{P}' \to A$, we have a homotopy equivalence $\tau : \mathbf{P} \approx \mathbf{P}'$, unique up to homotopy, and then

$$\tau \otimes B : \mathbf{P} \otimes_R B \approx \mathbf{P}' \otimes_R B$$

(since $- \otimes_R B$ is additive), so that there is a canonical isomorphism $H_n(\mathbf{P} \otimes_R B) \cong H_n(\mathbf{P}' \otimes_R B)$. We then argue as in Theorem 3.1 to infer

THEOREM 5.1. $\text{Tor}_n^R(-, -)$ *is an additive bifunctor, covariant in both variables.*

Continuing as in Section 3, we next prove

THEOREM 5.2.

$$\text{Tor}_0^R(A, B) \cong A \otimes_R B.$$

PROOF. As in the proof of Theorem 3.2, we start with the right exact sequence $P_1 \overset{\partial_1}{\to} P_0 \overset{\phi_0}{\to} A \to 0$. Since $- \otimes_R B$ is right exact, it follows that

$$P_1 \otimes_R B \overset{\partial_1}{\to} P_0 \otimes_R B \to A \otimes_R B \to 0$$

is right exact, so that $\text{Tor}_0^R(A, B) = \text{coker}\, \partial_1 \cong A \otimes_R B$.

THEOREM 5.3.

$$\text{Tor}_n^R(A, B) \cong \ker i_* : B_{n-1} \otimes_R B \to P_{n-1} \otimes_R B, \quad n \geq 1,$$

where

$$B_{n-1} = B_{n-1}(\mathbf{P}), \quad \text{and} \quad i : B_{n-1} \subseteq P_{n-1}$$

PROOF. As in the proof of Theorem 3.3, we apply the right exact functor $- \otimes_R B$ to the right exact sequence $P_{n+1} \overset{\partial_{n+1}}{\to} P_n \overset{\bar{\partial}}{\to} B_{n-1} \to 0$ to obtain

$$P_{n+1} \otimes_R B \overset{\partial_{n+1}}{\to} P_n \otimes_R B \overset{\bar{\partial}_*}{\to} B_{n-1} \otimes_R B \to 0, \tag{5.5}$$

where $\partial_n = i_* \bar{\partial}_* : P_n \otimes_R B \to P_{n-1} \otimes_R B$. Then $\bar{\partial}_*$ maps $\ker \partial_n$ onto $\ker i_*$, with kernel $\text{im}\, \partial_{n+1}$, whence the theorem follows.

THEOREM 5.4.

$$\text{Tor}_n^R(A, B) \cong \text{Tor}_{n-k}^R(B_{k-1}, B), \quad n \geq 2, 1 \leq k \leq n-1.$$

We leave the proof as an exercise.

We will not trouble to list explicitly the corresponding statements for $\overline{\text{Tor}}$.

We now pass to the exact sequences corresponding to those of Theorems 4.3 and 4.4.

THEOREM 5.5. *Let* $0 \to B' \overset{\mu}{\to} B \overset{\epsilon}{\to} B'' \to 0$ *be a short exact sequence of left R-modules. Then there is an exact sequence*

$$\cdots \to \text{Tor}_{n+1}^R(A,B'') \overset{\omega}{\to} \text{Tor}_n^R(A,B') \overset{\mu_*}{\to} \text{Tor}_n^R(A,B) \overset{\epsilon_*}{\to} \text{Tor}_n^R(A,B'') \to \cdots$$

$$\to \text{Tor}_1^R(A,B'') \overset{\omega}{\to} A \otimes_R B' \overset{\mu_*}{\to} A \otimes_R B \overset{\epsilon_*}{\to} A \otimes_R B'' \to 0.$$

PROOF.   Let $\mathbf{P} \to A$ be a projective resolution of A. Since each P_n is projective, it follows that the sequence

$$0 \to \mathbf{P} \otimes_R B' \overset{\mu_*}{\to} \mathbf{P} \otimes_R B \overset{\epsilon_*}{\to} \mathbf{P} \otimes_R B'' \to 0 \qquad (5.6)$$

is also short exact. We apply Theorem 4.2 to obtain the theorem (in view of Theorem 5.2).

THEOREM 5.6. *Let* $0 \to A' \overset{\mu}{\to} A \overset{\epsilon}{\to} A'' \to 0$ *be a short-exact sequence of right R-modules. Then there is an exact sequence*

$$\cdots \to \text{Tor}_{n+1}^R(A'',B) \overset{\omega}{\to} \text{Tor}_n^R(A',B) \overset{\mu_*}{\to} \text{Tor}_n^R(A,B) \overset{\epsilon_*}{\to} \text{Tor}_n^R(A'',B) \to \cdots$$

$$\to \text{Tor}_1^R(A'',B) \overset{\omega}{\to} A' \otimes_R B \overset{\mu_*}{\to} A \otimes_R B \overset{\epsilon_*}{\to} A'' \otimes_R B \to 0.$$

PROOF.   We apply Lemma 4.5 to find the commutative diagram (4.6). Just as in the proof of Theorem 4.4, we now deduce that

$$0 \to \mathbf{P}' \otimes_R B \overset{\mu_*}{\to} \mathbf{P} \otimes_R B \overset{\epsilon_*}{\to} \mathbf{P}'' \otimes_R B \to 0 \qquad (5.7)$$

is short exact, and obtain the theorem by applying Theorem 4.2.

We will prove analogs of Theorems 4.6 and 4.7 but it will be convenient to prove first the analog of Theorem 4.8. To this end, we are content to observe the following elementary proposition.

PROPOSITION 5.7.

   (*i*) *If* A *is projective, then* $Tor_n^R(A,B) = 0$ *for all* B *and all* $n \geqslant 1$.

(ii) If B *is projective, then* $Tor_n^R(A, B) = 0$ *for all* A *and all* $n \geqslant 1$.

PROOF.

(i) Choose a resolution **P** of A with $P_0 = A$, $P_n = 0, n \geqslant 1$.

(ii) If B is projective, $- \otimes_R B$ is exact, so that $H_n(\mathbf{P} \otimes_R B) = 0, n \geqslant 1$.

THEOREM 5.8.

$$Tor_n^R(A, B) \cong \overline{Tor}\,_n^R(A, B).$$

PROOF. For $n = 0$, each is isomorphic to $A \otimes_R B$. For $n = 1$, we take a projective presentation of B,

$$B_0 \overset{i}{\rightarrowtail} Q_0 \twoheadrightarrow B,$$

and apply Theorem 5.5 and Proposition 5.7(ii) to infer an exact sequence

$$0 \to Tor_1^R(A, B) \to A \otimes_R B_0 \overset{i_*}{\to} A \otimes_R Q_0 \to A \otimes_R B \to 0.$$

Thus $Tor_1^R(A, B) \cong \ker i_*$. But, by Theorem 5.3*, $\overline{Tor}\,_1^R(A, B) \cong \ker i_*$.
For $n \geqslant 2$, we argue by induction. We have

$$\overline{Tor}\,_n^R(A, B) \cong \overline{Tor}\,_{n-1}^R(A, B_0), \quad \text{by Theorem 5.4}^*,$$

$$\cong Tor_{n-1}^R(A, B_0), \quad \text{by the inductive hypothesis,}$$

$$\cong Tor_n^R(A, B),$$

since

$$\cdots \to Tor_n^R(A, Q_0) \to Tor_n^R(A, B) \to Tor_{n-1}^R(A, B_0) \to Tor_{n-1}^R(A, Q_0) \to \cdots$$

is exact and $Tor_n^R(A, Q_0) = Tor_{n-1}^R(A, Q_0) = 0$ by Proposition 5.7(ii).

REMARK. Once again, what is really involved here is a natural equivalence of bifunctors.

It is now not necessary to enunciate results in "dual" form, since plainly any statement relating to the A variable is matched by a corresponding statement relating to the B variable.

THEOREM 5.9. *The following statements about the right* R-*module* A *are equivalent*:

(*i*) A *is flat.*

(*ii*) $Tor_n^R(A, B) = 0$ *for all* B *and all* $n \geqslant 1$.

(*iii*) $Tor_1^R(A, B) = 0$ *for all* B.

PROOF.

(i)$\Rightarrow$(ii) Since A is flat, $A \otimes_R -$ is exact, so that $H_n(A \otimes_R \mathbf{Q}) = 0, n \geqslant 1$, where $\mathbf{Q}$ is a resolution of B.

(ii)$\Rightarrow$(iii) Trivial.

(iii)$\Rightarrow$(i) Let $B' \rightarrowtail B \twoheadrightarrow B''$ be exact. Then $\mathrm{Tor}_1^R(A, B'') \rightarrow A \otimes_R B' \rightarrow A \otimes_R B \rightarrow A \otimes_R B'' \rightarrow 0$ is exact. But $\mathrm{Tor}_1^R(A, B'') = 0$, so that $A \otimes_R -$ is exact and A is flat.

We do not prove any analog of Theorem 4.9. However, we close by proving a theorem establishing a link between Tor and Ext.

THEOREM 5.10. *The collection of natural transformations from the functor* $Ext_R^n(A, -) : \mathfrak{M}_R \rightarrow \mathfrak{Ab}$ *to the functor* $- \otimes_R B : \mathfrak{M}_R \rightarrow \mathfrak{Ab}$ *is in one-one correspondence with the elements of Tor* $_n^R(A, B)$.

PROOF.   In the case of $n = 0$ this is merely a special case of the Yoneda lemma (Exercise 3.4 of Chapter 3); hence we suppose that $n \geqslant 1$. Let $\mathbf{P} \rightarrow A$ be a projective resolution of A with $B_i = B_i(\mathbf{P}), i \geqslant 0$.

If $\tau : \mathrm{Ext}_R^n(A, -) \rightarrow - \otimes_R B$ is a natural transformation, consider $\tau_{B_{n-1}}$,

$$\tau_{B_{n-1}} : \mathrm{Ext}_R^n(A, B_{n-1}) \rightarrow B_{n-1} \otimes_R B.$$

Now, by Theorem 3.3. $\mathrm{Ext}_R^n(A, B_{n-1})$ contains the class of the identity map $1_{B_{n-1}}$. Thus $\tau[1_{B_{n-1}}] = t \in B_{n-1} \otimes_R B$. Consider now the embedding $i : B_{n-1} \rightarrow P_{n-1}$. We have a commutative diagram

$$
\begin{array}{ccc}
\mathrm{Ext}_R^n(A, B_{n-1}) & \overset{\tau}{\longrightarrow} & B_{n-1} \otimes_R B \\
\downarrow{i_*} & & \downarrow{i_*} \\
\mathrm{Ext}_R^n(A, P_{n-1}) & \overset{\tau}{\longrightarrow} & P_{n-1} \otimes_R B
\end{array}
\qquad (5.8)
$$

Moreover $i_*[1_{B_{n-1}}] = 0$, since $i : B_{n-1} \rightarrow P_{n-1}$ certainly extends to P_{n-1}. Thus (5.8) implies that $i_*(t) = 0$, or $t \in \mathrm{Tor}_n^R(A, B)$, by Theorem 5.3. Thus we may associate $t \in \mathrm{Tor}_n^R(A, B)$ with τ. We claim that this association is a one-one correspondence.

To see this, we start with $t \in \mathrm{Tor}_n^R(A, B)$ and proceed to build the corresponding τ. Of course, we set $\tau[1_{B_{n-1}}] = t$. We claim that τ is thereby completely determined by the requirement that it be natural. For let $\gamma \in \mathrm{Ext}_R^n(A, M)$. Then γ is represented by $\phi : B_{n-1} \rightarrow M$. The naturality of τ requires the commutativity of

$$
\begin{array}{ccc}
\mathrm{Ext}_R^n(A, B_{n-1}) & \overset{\tau}{\longrightarrow} & B_{n-1} \otimes_R B \\
\downarrow{\phi_*} & & \downarrow{\phi_*} \\
\mathrm{Ext}_R^n(A, M) & \overset{\tau}{\longrightarrow} & M \otimes_R B
\end{array}
$$

so that, since $\phi_*[1_{B_{n-1}}]=[\phi]=\gamma$, we are compelled to set

$$\tau(\gamma)=\phi_*(t). \tag{5.9}$$

Moreover, (5.9) is an unambiguous definition. For if ϕ' also represents γ, then $\phi'=\phi+\psi i$, $\psi:P_{n-1}\to M$, so that $\phi'_*(t)=\phi_*(t)+\psi_*i_*(t)=\phi_*(t)$, since $t\in\text{Tor}_n^R(A,B)$, hence $i_*(t)=0$.

It only remains to show that (5.9) does indeed specify a natural transformation. Let $\theta:M\to N$ and let $\gamma\in\text{Ext}_R^n(A,M)$ be represented by $\phi:B_{n-1}\to M$. Then $\theta_*(\gamma)$ is represented by $\theta\phi$, so that, according to (5.9),

$$\tau_N\theta_*(\gamma)=(\theta\phi)_*(t)=\theta_*\phi_*(t)=\theta_*\tau_M(\gamma).$$

Thus the theorem is completely proved.

6. PULLBACKS AND PUSHOUTS IN $_R\mathfrak{M}$

This section is to be regarded as preparatory for Section 7; here we record certain special features of pullbacks and pushouts (Section 6 of Chapter 3) present in $_R\mathfrak{M}$; indeed, all the remarks of this section generalize to any abelian category. There was some discussion of the special case of the category $\mathfrak{Ab}(R=\mathbf{Z})$ in Chapter 3 (Section 6).

Suppose that we are given a square of morphisms of $_R\mathfrak{M}$,

$$
\begin{array}{ccc}
A_0 & \xrightarrow{\phi_1} & A_1 \\
{\scriptstyle\phi_2}\downarrow & & \downarrow{\scriptstyle\psi_1} \\
A_2 & \xrightarrow{\psi_2} & A
\end{array}
\tag{6.1}
$$

With (6.1) we may associate the sequence of morphisms of $_R\mathfrak{M}$,

$$A_0 \xrightarrow{\{\phi_1,\phi_2\}} A_1\oplus A_2 \xrightarrow{\langle\psi_1,-\psi_2\rangle} A \tag{6.2}$$

Now the composite morphism in (6.2) is $\psi_1\phi_1-\psi_2\phi_2$. Thus we see

PROPOSITION 6.1. *The square* (6.1) *commutes if and only if the sequence* (6.2) *is differential* (*that is, the composite morphism is zero*).

Note. In (6.2) we took ψ_2 with the minus sign. All that matters is that we take *one* of the morphisms of (6.1) with the minus sign. None of our results depends on our singling out ψ_2 for this honor!

We now establish the vital connection between (6.1) and (6.2).

THEOREM 6.2.

(i) *The square* (6.1) *is a pullback if and only if* $\{\phi_1, \phi_2\}$ *is the kernel of* $\langle \psi_1, -\psi_2 \rangle$ *in* (6.2).

(ii) *The square* (6.1) *is a pushout if and only if* $\langle \psi_1, -\psi_2 \rangle$ *is the cokernel of* $\{\phi_1, \phi_2\}$ *in* (6.2).

PROOF. We will be content to prove (i). We may identify a pair of morphisms $\xi_i : X \to A_i, i = 1, 2$, with the morphism $\{\xi_1, \xi_2\} : X \to A_1 \oplus A_2$. Then the following assertion is clear: (6.1) is a pullback⟺for all $\{\xi_1, \xi_2\} : X \to A_1 \oplus A_2$ such that $\langle \psi_1, -\psi_2 \rangle \{\xi_1, \xi_2\} = 0$, there exists a unique $\xi : X \to A_0$, such that $\{\phi_1, \phi_2\}\xi = \{\xi_1, \xi_2\}$.

But this is precisely the (categorical) description of the fact that $\{\phi_1, \phi_2\}$ is the kernel of $\langle \psi_1, -\psi_2 \rangle$.

DEFINITION 6.1. We say that (6.1) is *bicartesian* if it is a pullback and a pushout.

COROLLARY 6.3.

(i) *Suppose that* (6.1) *is a pullback and* $\langle \psi_1, -\psi_2 \rangle$ *is epic. Then* (6.1) *is bicartesian.*

(ii) *Suppose that* (6.1) *is a pushout and* $\{\phi_1, \phi_2\}$ *is monic. Then* (6.1) *is bicartesian.*

PROOF. (i) If $\{\phi_1, \phi_2\}$ is the kernel of $\langle \psi_1, -\psi_2 \rangle$ and $\langle \psi_1, -\psi_2 \rangle$ is epic, then $\langle \psi_1, -\psi_2 \rangle$ is the cokernel of $\{\phi_1, \phi_2\}$. Similarly we prove (ii).

REMARK. Of course, $\langle \psi_1, -\psi_2 \rangle$ is epic if one of ψ_1, ψ_2 is epic. It is often the case that we apply Corollary 6.3(i) when one of ψ_1, ψ_2 is epic, and Corollary 6.3(ii) when one of ϕ_1, ϕ_2 is monic.

The statements that follow have duals, which we will enunciate but not prove.

THEOREM 6.4. *Let* (6.1) *be a pullback in* $_R\mathfrak{M}$. *Then the induced map of kernels,* $\ker \phi_1 \to \ker \psi_2$, *is an isomorphism. Conversely, let* (6.1) *be a commutative square in* $_R\mathfrak{M}$ *such that the induced map* $\ker \phi_1 \to \ker \psi_2$ *is an isomorphism. Then, provided that* ϕ_1 *is surjective,* (6.1) *is a pullback.*

PROOF. The first statement is true in any category. For the converse, we must show that $\{\phi_1, \phi_2\}$ is monic and that, given $(a_1, a_2) \in A_1 \oplus A_2$ with $\psi_1 a_1 = \psi_2 a_2$, there exists $a_0 \in A_0$ with $\phi_1 a_0 = a_1, \phi_2 a_0 = a_2$.

First, let $\phi_1 a_0 = 0$, $\phi_2 a_0 = 0$. Then $a_0 \in \ker \phi_1$ and, as ϕ_2 maps $\ker \phi_1$ isomorphically onto $\ker \psi_2$, it follows that $a_0 = 0$. Second, given $(a_1, a_2) \in A_1 \oplus A_2$ with $\psi_1 a_1 = \psi_2 a_2$, we first put $a_1 = \phi_1 b_0$, $b_0 \in A_0$—this is legitimate, since ϕ_1 is surjective. Then $\psi_2 a_2 = \psi_1 \phi_1 b_0 = \psi_2 \phi_2 b_0$, so that $a_2 = \phi_2 b_0 + c_2, c_2 \in \ker \psi_2$. Since ϕ_2 maps $\ker \phi_1$ onto $\ker \psi_2, c_2 = \phi_2 c_0$,

$c_0 \in \ker \phi_1$. Set $a_0 = b_0 + c_0$. Then $a_1 = \phi_1 b_0 = \phi_1 a_0$ and $a_2 = \phi_2 a_0$, so that the theorem is proved.

THEOREM 6.4*. *Let* (6.1) *be a pushout in* $_R\mathfrak{M}$. *Then the induced map of cokernels* $\operatorname{coker} \phi_1 \to \operatorname{coker} \psi_2$ *is an isomorphism. Conversely, let* (6.1) *be a commutative square in* $_R\mathfrak{M}$ *such that the induced map* $\operatorname{coker} \phi_1 \to \operatorname{coker} \psi_2$ *is an isomorphism. Then, provided that* ψ_2 *is injective,* (6.1) *is a pushout.*

THEOREM 6.5. *Let* (6.1) *be a pullback. Then if* ψ_2 *is surjective,* ϕ_2 *is surjective.*

PROOF. By Corollary 6.3(i), (6.1) is bicartesian. But, in any category, (6.1) being a pushout, ϕ_1 and ψ_2 have isomorphic cokernels.

THEOREM 6.5*. *Let* (6.1) *be a pushout. Then if* ϕ_1 *is injective,* ψ_2 *is injective.*

Now let A, B be fixed R-modules and let

$$0 \to B \xrightarrow{\mu} E \xrightarrow{\epsilon} A \to 0$$

be a short exact sequence. We introduce an equivalence relation into such short exact sequences by declaring that $0 \to B \xrightarrow{\mu} E \xrightarrow{\epsilon} A \to 0$ and $0 \to B \xrightarrow{\mu'} E' \xrightarrow{\epsilon'} A \to 0$ are equivalent if there exists an isomorphism $\eta: E \to E'$ which reduces to the identity on A and B; that is, if there is a commutative diagram

$$
\begin{array}{ccccc}
B & \rightarrowtail & E & \twoheadrightarrow & A \\
\| & & \downarrow{\scriptstyle \eta} & & \| \\
B & \xrightarrow{\mu'} & E' & \xrightarrow{\epsilon'} & A'
\end{array}
\qquad (6.3)
$$

It is an easy consequence of Theorem 4.1 that any η making the diagram (6.3) commutative must indeed be an isomorphism. We write **E** for the equivalence class of $B \rightarrowtail E \twoheadrightarrow A$ and call **E** an *extension* of A by B. We may write

$$\mathbf{E}: B \xrightarrow{\mu} E \xrightarrow{\epsilon} A \qquad (6.4)$$

for an extension represented by the given short exact sequence.

We consider the extension (6.4) and let $\alpha: A_1 \to A, \beta: B \to B_1$ be homomorphisms. Then we may form the pullback of α and ϵ,

$$
\begin{array}{ccc}
E_\alpha & \xrightarrow{\epsilon_\alpha} & A_1 \\
\downarrow{\scriptstyle \rho} & & \downarrow{\scriptstyle \alpha} \\
E & \xrightarrow{\ \epsilon\ } & A
\end{array}
$$

By Theorem 6.5, ϵ_α is epic, and, by Theorem 6.4, the kernel of ϵ_α is B. Thus we have a map of short exact sequences,

$$
\begin{array}{ccccc}
\mathbf{E}_\alpha: & B & \xrightarrow{\mu_\alpha} & E & \xrightarrow{\epsilon_\alpha} & A_1 \\
& \downarrow & & \Big\| & & \downarrow{\scriptstyle \rho} \qquad \downarrow{\scriptstyle \alpha} \\
\mathbf{E}: & B & \xrightarrow{\ \mu\ } & E & \xrightarrow{\ \epsilon\ } & A
\end{array} \tag{6.5}
$$

and we write $\mathbf{E}_\alpha = \alpha^*(\mathbf{E})$. This is well-defined, as the reader will readily verify.

Similarly, we may form the pushout of β and μ,

$$
\begin{array}{ccc}
B & \xrightarrow{\mu} & E \\
{\scriptstyle \beta}\downarrow & & \downarrow{\scriptstyle \sigma} \\
B_1 & \xrightarrow{\beta\mu} & {}_\beta E
\end{array}
$$

and, utilizing Theorems 6.4* and 6.5*, we obtain a map of short exact sequences

$$
\begin{array}{ccccc}
\mathbf{E}: & B & \xrightarrow{\mu} & E & \xrightarrow{\epsilon} & A \\
& \downarrow & \downarrow{\scriptstyle \beta} & \downarrow{\scriptstyle \sigma} & & \Big\| \\
{}_\beta\mathbf{E}: & B_1 & \xrightarrow{\beta\mu} & {}_\beta E & \xrightarrow{\beta\epsilon} & A
\end{array} \tag{6.6}
$$

and we write ${}_\beta\mathbf{E} = \beta_*(\mathbf{E})$.

THEOREM 6.6.
    (i) If $\alpha : A_1 \to A, \alpha' : A_2 \to A_1$, then $(\alpha\alpha')^* = \alpha'^*\alpha^*$.
    (ii) If $\beta : B \to B_1, \beta' : B_1 \to B_2$, then $(\beta'\beta)_* = \beta'_*\beta_*$.

PROOF.

(i) We merely observe (Exercise 6.1 of Chapter 3) that the composite of two pullback squares is a pullback square.

(ii) Dually, the composite of two pushout squares is a pushout square.

THEOREM 6.7. *Let* $E: B \rightarrowtail_2 E \twoheadrightarrow A$ *be an extension and let* $\alpha: A_1 \to A, \beta: B \to B_1$. *Then*

$$\beta_*\alpha^*E = \alpha^*\beta_*E$$

PROOF. We have the two diagrams

$$
\begin{array}{ccccccc}
\beta_*\alpha^*E: & B_1 & \xrightarrow{\;\beta(\mu_\alpha)\;} & {}_\beta(E_\alpha) & \xrightarrow{\;\beta(\epsilon_\alpha)\;} & A_1 \\[4pt]
 & \Big\uparrow{\scriptstyle\beta} & \text{Pushout} & \Big\uparrow{\scriptstyle\sigma'} & & \Big\| \\[4pt]
\alpha^*E: & B & \xrightarrow{\;\mu_\alpha\;} & E_\alpha & \xrightarrow{\;\epsilon_\alpha\;} & A_1 & \qquad (6.7)\\[4pt]
 & \Big\| & & \Big\downarrow{\scriptstyle\rho} & \text{Pullback} & \Big\downarrow{\scriptstyle\alpha} \\[4pt]
E: & B & \xrightarrow{\;\mu\;} & E & \xrightarrow{\;\epsilon\;} & A
\end{array}
$$

$$
\begin{array}{ccccccc}
\alpha^*\beta_*E: & B_1 & \xrightarrow{\;(\beta\mu)_\alpha\;} & (_\beta E)_\alpha & \xrightarrow{\;(\beta\epsilon)_\alpha\;} & A_1 \\[4pt]
 & \Big\| & & \Big\downarrow{\scriptstyle\rho'} & \text{Pullback} & \Big\downarrow{\scriptstyle\alpha} \\[4pt]
\beta_*E: & B_1 & \xrightarrow{\;\beta\mu\;} & _\beta E & \xrightarrow{\;\beta\epsilon\;} & A & \qquad (6.8)\\[4pt]
 & \Big\uparrow{\scriptstyle\beta} & \text{Pushout} & \Big\uparrow{\scriptstyle\sigma} & & \Big\| \\[4pt]
E: & B & \xrightarrow{\;\mu\;} & E & \xrightarrow{\;\epsilon\;} & A
\end{array}
$$

and we must obtain a homomorphism $\eta: {}_\beta(E_\alpha) \to (_\beta E)_\alpha$ such that the diagram

$$
\begin{array}{ccccc}
B_1 & \xrightarrow{\;\beta(\mu_\alpha)\;} & {}_\beta(E_\alpha) & \xrightarrow{\;\beta(\epsilon_\alpha)\;} & A_1 \\[4pt]
\Big\| & & \Big\downarrow{\scriptstyle\eta} & & \Big\| \\[4pt]
B_1 & \xrightarrow{\;(\beta\mu)_\alpha\;} & (_\beta E)_\alpha & \xrightarrow{\;(\beta\epsilon)_\alpha\;} & A_1
\end{array} \qquad (6.9)
$$

commutes. We will argue from the categorical definitions of pullback and pushout. First we obtain a map $\theta : E_\alpha \rightarrow (_\beta E)_\alpha$ such that

$$\rho'\theta = \sigma\rho, \quad (_\beta\epsilon)_\alpha\theta = \epsilon_\alpha \qquad (6.10)$$

Such a map θ exists (and is unique), since we have a pullback in (6.8) and

$$(_\beta\epsilon)\sigma\rho = \epsilon\rho = \alpha\epsilon_\alpha.$$

We next argue that

$$\theta\mu_\alpha = (_\beta\mu)_\alpha\beta. \qquad (6.11)$$

For

$$\rho'\theta\mu_\alpha = \sigma\rho\mu_\alpha = \sigma\mu = (_\beta\mu)\beta = \rho'(_\beta\mu)_\alpha\beta;$$

and

$$(_\beta\epsilon)_\alpha\theta\mu_\alpha = \epsilon_\alpha\mu_\alpha = 0 = (_\beta\epsilon)_\alpha(_\beta\mu)_\alpha\beta.$$

Thus, from the pushout in (6.7), we infer $\eta : _\beta(E_\alpha) \rightarrow (_\beta E)_\alpha$ such that

$$\eta_\beta(\mu_\alpha) = (_\beta\mu)_\alpha, \quad \eta\sigma' = \theta. \qquad (6.12)$$

It remains to show that

$$(_\beta\epsilon)_\alpha\eta = _\beta(\epsilon_\alpha). \qquad (6.13)$$

But $(_\beta\epsilon)_\alpha\eta_\beta(\mu_\alpha) = (_\beta\epsilon)_\alpha(_\beta\mu)_\alpha = 0 = _\beta(\epsilon_\alpha)_\beta(\mu_\alpha)$; and

$$(_\beta\epsilon)_\alpha\eta\sigma' = (_\beta\epsilon)_\alpha\theta = \epsilon_\alpha = _\beta(\epsilon_\alpha)\sigma'.$$

Thus (6.13) is established and the proof of the theorem is complete.

We prove one more theorem related to extensions. Given two extensions

$$\mathbf{E}_i : \qquad\qquad B_i \overset{\mu_i}{\rightarrowtail} E_i \overset{\epsilon_i}{\twoheadrightarrow} A_i, \qquad i = 1, 2,$$

we may form their *direct sum*

$$\mathbf{E}_1 \oplus \mathbf{E}_2 : \qquad\qquad B_1 \oplus B_2 \overset{\mu_1 \oplus \mu_2}{\rightarrowtail} E_1 \oplus E_2 \overset{\epsilon_1 \oplus \epsilon_2}{\twoheadrightarrow} A_1 \oplus A_2.$$

This construction will play a prominent role in the next section. Here we prove just one basic result relating to it.

THEOREM 6.8. *Let* $\mathbf{E} : B \rightarrowtail E \twoheadrightarrow A$ *be an extension and let* $\Delta_A : A \rightarrow A \oplus A, \Delta_B : B \rightarrow B \oplus B$ *be diagonal maps. Then* $\Delta_{B*}\mathbf{E} = \Delta_A^*(\mathbf{E} \oplus \mathbf{E})$.

PROOF. We have diagrams

$$
\begin{array}{ccccc}
B & \xrightarrow{\ \mu\ } & E & \xrightarrow{\ \epsilon\ } & A \\[4pt]
\Big\downarrow{\scriptstyle \Delta_B} & \text{Pushout} & \Big\downarrow{\scriptstyle \sigma} & & \Big\| \\[4pt]
B \oplus B & \xrightarrow{\ \mu_0\ } & E_0 & \xrightarrow{\ \epsilon_0\ } & A
\end{array}
\qquad (6.14)
$$

$$
\begin{array}{ccccc}
B \oplus B & \xrightarrow{\ \mu \oplus \mu\ } & E \oplus E & \xrightarrow{\ \epsilon \oplus \epsilon\ } & A \oplus A \\[4pt]
\Big\| & & \Big\uparrow{\scriptstyle \sigma'} & \text{Pullback} & \Big\uparrow{\scriptstyle \Delta_A} \\[4pt]
B \oplus B & \xrightarrow{\ \mu_1\ } & E_1 & \xrightarrow{\ \epsilon_1\ } & A
\end{array}
\qquad (6.15)
$$

and our objective is to construct $\eta : E_0 \to E_1$ with

$$\eta\mu_0 = \mu_1, \quad \epsilon_1\eta = \epsilon_0. \qquad (6.16)$$

Our argument is similar to that of Theorem 6.7, but simpler. We first construct $\theta : E \to E_1$ satisfying

$$\sigma'\theta = \Delta_E, \quad \epsilon_1\theta = \epsilon. \qquad (6.17)$$

This is possible since we have a pullback in (6.15) and

$$(\epsilon \oplus \epsilon)\Delta = \Delta\epsilon.$$

We then argue that

$$\theta\mu = \mu_1\Delta_B. \qquad (6.18)$$

For $\sigma'\theta\mu = \Delta\mu = (\mu \oplus \mu)\Delta = \sigma'\mu_1\Delta$; and $\epsilon_1\theta\mu = \epsilon\mu = 0 = \epsilon_1\mu_1\Delta$. Thus from the pushout in (6.14) we infer $\eta : E_0 \to E_1$ with

$$\eta\mu_0 = \mu_1, \quad \eta\sigma = \theta. \qquad (6.19)$$

To infer, finally, that $\epsilon_1\eta = \epsilon_0$, it suffices to observe that

$$\epsilon_1\eta\mu_0 = \epsilon_1\mu_1 = 0 = \epsilon_0\mu_0;$$

and

$$\epsilon_1\eta\sigma = \epsilon_1\theta = \epsilon = \epsilon_0\sigma.$$

This completes the proof.

7. Ext^1 AND EXTENSIONS

In this section we set up a one-one correspondence between elements of $\mathrm{Ext}_R^1(A,B)$ and elements of the set $E(A,B)$ of *extensions* $\mathbf{E}$ of the quotient module A by the submodule B. This correspondence serves, indeed, to explain and justify the choice of terminology Ext. There is, actually, an analog for Ext^n, but we will be content to lead the reader through this in the Exercises.

In fact, we establish more than a one-one correspondence between elements. For we may summarize Theorems 6.6 and 6.7 (together with the trivial relations $1_* = 1, 1^* = 1$) as asserting

PROPOSITION 7.1. $E(-,-)$ *is a bifunctor* $_R\mathfrak{M}^0 \times {}_R\mathfrak{M} \rightarrow \mathfrak{S}$, *the category of sets.*

We will show that, if, temporarily, we also regard $\mathrm{Ext}_R^1(-,-)$ as a bifunctor to sets (that is, speaking more strictly, if we compose $\mathrm{Ext}_R^1(-,-)$ with the underlying functor to sets), then $E(-,-)$ and $\mathrm{Ext}_R^1(-,-)$ are naturally equivalent. We now describe the natural equivalence.

We start with a fixed projective presentation

$$Q \overset{\nu}{\rightarrowtail} P \overset{\epsilon}{\twoheadrightarrow} A \tag{7.1}$$

of A. We recall that, according to Theorem 3.3, there is then a natural identification

$$\mathrm{Ext}_R^1(A,B) = \mathrm{Hom}_R(Q,B)/\nu^*\mathrm{Hom}_R(P,B). \tag{7.2}$$

Now let

$$\mathbf{E}: B \overset{\mu}{\rightarrowtail} E \overset{\epsilon}{\twoheadrightarrow} A$$

be an extension. Since P is projective, we may construct a commutative diagram

$$
\begin{array}{ccccc}
Q & \overset{\nu}{\rightarrowtail} & P & \overset{\eta}{\twoheadrightarrow} & A \\
\downarrow{\psi} & & \downarrow{\phi} & & \parallel \\
B & \overset{\mu}{\rightarrowtail} & E & \overset{\epsilon}{\twoheadrightarrow} & A
\end{array}
\tag{7.3}
$$

Now, in (7.3), we may alter ϕ by adding a homomorphism $\mu\kappa$, where κ is an arbitrary homomorphism $\kappa: P \rightarrow B$. We would then alter ψ by adding $\kappa\nu$. It follows that ψ is determined *modulo* $\nu^*\mathrm{Hom}_R(P,B)$, so that we have determined a perfectly definite element $\zeta = [\psi]$ of $\mathrm{Hom}_R(Q,B)/\nu^*\mathrm{Hom}_R(P,B)$.

According to (7.2), we may regard ζ as an element of $\text{Ext}_R^1(A,B)$. It is plain that if we replace $B \rightarrowtail E \twoheadrightarrow A$ by another short exact sequence also representing $\mathbf{E}$ [see (6.3)], we obtain the same element ξ. Thus the rule $\mathbf{E} \mapsto \xi$ establishes a function

$$\theta : E(A,B) \rightarrow \text{Ext}_R^1(A,B).$$

THEOREM 7.2. θ *is a natural equivalence of bifunctors to sets.*

PROOF. We first establish that θ is a one-one correspondence. Suppose that we have $\xi \in \text{Ext}_R^1(A,B)$, represented by $\psi : Q \rightarrow B$ according to (7.2). Form the pushout of ψ and $\nu : Q \rightarrowtail P$,

$$
\begin{array}{ccc}
Q & \overset{\nu}{\rightarrowtail} & P \\
{\scriptstyle \psi}\downarrow & & \downarrow{\scriptstyle \phi} \\
B & \underset{\mu}{\rightarrowtail} & E
\end{array}
$$

Note that μ is monic by Theorem 6.5*, and that the induced map $\text{coker}\,\nu \rightarrow \text{coker}\,\mu$ is an isomorphism by Theorem 6.4*. Thus we may identify $\text{coker}\,\mu$ with A and we obtain the commutative diagram

$$
\begin{array}{ccccc}
Q & \overset{\nu}{\rightarrowtail} & P & \overset{\eta}{\twoheadrightarrow} & A \\
{\scriptstyle \psi}\downarrow & \text{Pushout} & \downarrow{\scriptstyle \phi} & & \| \\
B & \underset{\mu}{\rightarrowtail} & E & \underset{\epsilon}{\twoheadrightarrow} & A
\end{array}
\qquad (7.4)
$$

with exact rows. Suppose that we choose a new representative $\psi + \kappa\nu$ of ξ, where $\kappa : P \rightarrow B$. If we also change ϕ to $\phi + \mu\kappa$ we preserve commutativity in the diagram corresponding to (7.4) without altering the homomorphisms ν, η, μ, ϵ. It follows from the second assertion of Theorem 6.4* that the left-hand square in the modified diagram remains a pushout. Thus the extension $\mathbf{E} : B \rightarrowtail E \twoheadrightarrow A$, obtained in (7.4) by means of ψ, in fact only depends on ξ. It is now obvious that the rule

$$\xi \mapsto \mathbf{E},$$

given by associating with ψ, representing ξ, the extension containing the bottom row of (7.4), yields a function $\text{Ext}_R^1(A,B) \rightarrow E(A,B)$ inverse to θ.

It remains to prove naturality. First we consider naturality in B. Thus let $\beta : B_1 \to B_2$, and let

$$\mathbf{E}_1: \qquad\qquad B_1 \rightarrowtail E_1 \twoheadrightarrow A$$

belong to $E(A, B_1)$. Form the diagram [see (6.6) and (7.3)]

$$
\begin{array}{ccccc}
Q & \overset{\nu}{\rightarrowtail} & P & \overset{\eta}{\twoheadrightarrow} & A \\
\downarrow{\scriptstyle \psi} & & \downarrow{\scriptstyle \phi} & & \| \\
\mathbf{E}_1: \quad B_1 & \overset{\mu_1}{\rightarrowtail} & E_1 & \overset{\epsilon_1}{\dashrightarrow\!\!\!\rightarrow} & A \\
\downarrow{\scriptstyle \beta}\ \text{Pushout} & \downarrow{\scriptstyle \sigma} & & & \| \\
\mathbf{E}_2: \quad B_2 & \overset{\mu_2}{\rightarrowtail} & E_2 & \overset{\epsilon_2}{\twoheadrightarrow} & A .
\end{array}
$$

Then $\theta(\mathbf{E}_1)$ is the equivalence class ξ of ψ and $\beta_* \theta(\mathbf{E}_1)$ is the equivalence class $\beta_* \xi$ of $\beta\psi$. On the other hand $\beta_*(\mathbf{E}_1) = \mathbf{E}_2$, and $\theta(\mathbf{E}_2)$ is the equivalence class of $\beta\psi$. Thus $\beta_* \theta = \theta\beta_*$.

Second, we consider naturality in A. Thus let $\alpha : A_2 \to A_1$, and let

$$\mathbf{E}_1: \qquad\qquad B \rightarrowtail E_1 \twoheadrightarrow A_1$$

belong to $E(A_1, B)$. Given projective presentations

$$Q_i \overset{\nu_i}{\rightarrowtail} P_i \overset{\eta_i}{\twoheadrightarrow} A_i, \qquad i = 1, 2,$$

we may form the commutative diagram

$$
\begin{array}{ccccc}
Q_2 & \rightarrowtail & P_2 & \longrightarrow & A_2 \\
\downarrow{\scriptstyle \gamma} & & \downarrow{\scriptstyle \beta} & & \downarrow{\scriptstyle \alpha} \\
Q_1 & \rightarrowtail & P_1 & \longrightarrow & A_1,
\end{array}
\qquad (7.5)
$$

since P_2 is projective, and if $\psi : Q_1 \to B$ represents $\xi \in \operatorname{Ext}_R^1(A_1, B)$, then $\psi\gamma : Q_2 \to B$ represents $\alpha^* \xi \in \operatorname{Ext}_R^1(A_2, B)$. We may construct β and γ in (7.5) as we please, subject only to requiring commutativity. We choose to construct them as follows:

$$\mathbf{P}_2: \quad Q_2 \rightarrowtail P_2 \twoheadrightarrow A_2$$

$$\downarrow \gamma \qquad \downarrow \delta \qquad \|$$

$$\mathbf{S}: \quad Q_1 \rightarrowtail S \twoheadrightarrow A_2 \qquad \beta = \rho\delta \qquad (7.6)$$

$$\| \qquad \downarrow \rho \text{ Pullback } \downarrow \alpha$$

$$\mathbf{P}_1: \quad Q_1 \rightarrowtail P_1 \twoheadrightarrow A_1$$

where the exactness of the middle row follows from Theorems 6.4 and 6.5. Now we may interpret

$$\mathbf{P}_1: \quad Q_1 \rightarrowtail P_1 \twoheadrightarrow A_1$$

$$\downarrow \psi \qquad \downarrow \phi \qquad \| \qquad (7.7)$$

$$\mathbf{E}_1: \quad B \rightarrowtail E_1 \twoheadrightarrow A_1$$

as saying that $\mathbf{E}_1 = \psi_* \mathbf{P}_1$, since, by Theorem 6.4*, the left-hand square of (7.7) is a pushout; similarly, the bottom half of (7.6) says that $\mathbf{S} = \alpha^* \mathbf{P}_1$.

Now $\theta(\mathbf{E}_1)$ is represented by $\psi: Q_1 \rightarrow B$ and $\alpha^* \theta(\mathbf{E}_1)$ is represented by $\psi\gamma: Q_2 \rightarrow B$. On the other hand,

$$\alpha^* \mathbf{E}_1 = \alpha^* \psi_* \mathbf{P}_1 = \psi_* \alpha^* \mathbf{P}_1, \quad \text{by Theorem 6.7,}$$

$$= \psi_* \mathbf{S}.$$

Thus

$$\theta\alpha^* \mathbf{E}_1 = \theta(\psi_* \mathbf{S}) = \psi_* \theta(\mathbf{S}),$$

by what we have already proved. Since (7.6) $\theta(\mathbf{S})$ is the class of γ, it follows that $\psi_* \theta(\mathbf{S})$ is the class of $\psi\gamma$, so that the proof of naturality is complete.

It is, of course, now open to us to use θ to transfer all information about $\mathrm{Ext}_R^1(A, B)$ across to $E(A, B)$. In particular we may use θ to induce the structure of an abelian group into each set $E(A, B)$ so that $E(-, -)$ becomes an additive bifunctor to abelian groups. Since this procedure, while logically satisfactory, would leave the nature of the addition law very obscure, we prefer to describe the law explicitly and show how one could prove the facts mentioned without appeal to the natural equivalence θ.

Given extensions

$$\mathbf{E}_i: \qquad B \overset{\mu_i}{\rightarrowtail} E_i \overset{\epsilon_i}{\twoheadrightarrow} A,$$

we define their sum $\mathbf{E}_1 + \mathbf{E}_2$ by the rule

$$\mathbf{E}_1 + \mathbf{E}_2 = \Delta^* \nabla_* (\mathbf{E}_1 \oplus \mathbf{E}_2), \tag{7.8}$$

where $\Delta = \{1,1\} : A \to A \oplus A$ is the diagonal, $\nabla = \langle 1,1 \rangle : B \oplus B \to B$ is the codiagonal, and $\mathbf{E}_1 \oplus \mathbf{E}_2$ is the *direct sum* (see Section 6).

To show that this is the "right" definition we immediately establish

PROPOSITION 7.3.

$$\theta(\mathbf{E}_1 + \mathbf{E}_2) = \theta(\mathbf{E}_1) + \theta(\mathbf{E}_2).$$

PROOF.   Let $\theta(\mathbf{E}_i) = [\psi_i]$, where $\psi_i : Q \to B$. Thus we have commutative diagrams

$$
\begin{array}{ccccc}
Q & \rightarrowtail & P & \twoheadrightarrow & A \\
\downarrow{\scriptstyle \psi_i} & & \downarrow{\scriptstyle \phi_i} & & \| \qquad i = 1,2. \\
B & \rightarrowtail & E_i & \twoheadrightarrow & A
\end{array}
$$

Consider the diagram

$$
\begin{array}{ccccc}
Q & \rightarrowtail & P & \twoheadrightarrow & A \\
\downarrow{\scriptstyle \Delta} & & \downarrow{\scriptstyle \Delta} & & \downarrow{\scriptstyle \Delta} \\
Q \oplus Q & \rightarrowtail & P \oplus P & \twoheadrightarrow & A \oplus A \\
\downarrow{\scriptstyle \psi_1 \oplus \psi_2} & & \downarrow{\scriptstyle \phi_1 \oplus \phi_2} & & \| \\
B \oplus B & \rightarrowtail & E_1 \oplus E_2 & \twoheadrightarrow & A \oplus A
\end{array}
$$

The bottom part shows that $\theta(\mathbf{E}_1 \oplus \mathbf{E}_2) = [\psi_1 \oplus \psi_2]$, and the top part shows that $\Delta^*[\psi_1 \oplus \psi_2] = [(\psi_1 \oplus \psi_2)\Delta]$. Thus

$$\theta(\mathbf{E}_1 + \mathbf{E}_2) = \theta(\Delta^* \nabla_* (\mathbf{E}_1 \oplus \mathbf{E}_2)) = \Delta^* \nabla_* \theta(\mathbf{E}_1 \oplus \mathbf{E}_2), \quad \text{by Theorem 7.2,}$$

$$= \Delta^* \nabla_* [\psi_1 \oplus \psi_2]$$

$$= \nabla_* [(\psi_1 \oplus \psi_2)\Delta]$$

$$= [\nabla(\psi_1 \oplus \psi_2)\Delta]$$

$$= [\psi_1 + \psi_2]$$

$$= [\psi_1] + [\psi_2]$$

$$= \theta(\mathbf{E}_1) + \theta(\mathbf{E}_2).$$

Having proved this proposition, we may, as mentioned earlier, infer that the rule (7.8) converts $E(-,-)$ into an additive bifunctor to abelian groups, merely by applying θ. However, we demonstrate the facts needed without appeal to θ, in order to convey some of the flavor of categorical reasoning.

PROPOSITION 7.4.
(i) $\alpha^*(E_1 + E_2) = \alpha^*E_1 + \alpha^*E_2$.
(ii) $\beta_*(E_1 + E_2) = \beta_*E_1 + \beta_*E_2$.

PROOF. We will be content to prove (i). Since, obviously, pullbacks commute with (finite) direct sums, we have

$$(\alpha_1 \oplus \alpha_2)^*(E_1 \oplus E_2) = \alpha_1^*E \oplus \alpha_2^*E. \tag{7.9}$$

Since Δ is natural, we have

$$\Delta\alpha = (\alpha \oplus \alpha)\Delta. \tag{7.10}$$

Thus

$\alpha^*(E_1 + E_2)$

$= \alpha^*\Delta^*\nabla_*(E_1 \oplus E_2) = \Delta^*\nabla_*(\alpha \oplus \alpha)^*(E_1 \oplus E_2)$, by (7.10) and Theorem 6.7,

$= \Delta^*\nabla_*(\alpha^*E_1 \oplus \alpha^*E_2)$,   by (7.9),

$= \alpha^*E_1 + \alpha^*E_2$.

PROPOSITION 7.5.
(i) $(\alpha_1 + \alpha_2)^*E = \alpha_1^*E + \alpha_2^*E$.
(ii) $(\beta_1 + \beta_2)_*E = \beta_1{}_*E + \beta_2{}_*E$.

PROOF. Again we will be content to prove (i). We have $\alpha_1 + \alpha_2 = \nabla(\alpha_1 \oplus \alpha_2)\Delta$. Thus

$(\alpha_1 + \alpha_2)^*E = \Delta^*(\alpha_1 \oplus \alpha_2)^*\nabla^*E$

$= \Delta^*(\alpha_1 \oplus \alpha_2)^*\nabla_*(E \oplus E)$,   by the dual of Theorem 6.8,

$= \Delta^*\nabla_*(\alpha_1 \oplus \alpha_2)^*(E \oplus E)$,   by Theorem 6.7,

$= \Delta^*\nabla_*(\alpha_1^*E \oplus \alpha_2^*E)$,   by (7.9),

$= \alpha_1^*E + \alpha_2^*E$.

PROPOSITION 7.6. *The zero element of* $E(A,B)$ *is the split extension* $Z: B \rightarrowtail B \oplus A \twoheadrightarrow A$, *and* $0^*E = Z$, $0_*E = Z$.

PROOF. The first assertion is trivial if we use θ. We have only to construct the commutative diagram

$$
\begin{array}{ccccc}
Q & \overset{\nu}{\rightarrowtail} & P & \overset{\eta}{\twoheadrightarrow} & A \\
\downarrow^{0} & & \downarrow^{\{0,\eta\}} & & \| \\
B & \rightarrowtail & B \oplus A & \twoheadrightarrow & A
\end{array}
$$

Now the commutative diagram

$$
\begin{array}{ccccc}
B & \rightarrowtail & B \oplus A_1 & \twoheadrightarrow & A_1 \\
\| & & \downarrow^{\langle \mu, 0 \rangle} & & \downarrow^{0} \\
B & \overset{\mu}{\rightarrowtail} & E & \overset{\epsilon}{\twoheadrightarrow} & A
\end{array}
$$

shows that $0^*E = Z$. Similarly, $0_*E = Z$. Note that we may now prove that the split extension Z is the zero element without appeal to θ as follows. We have proved that $0^*E = Z$. Thus

$$E = 1^*E = (1+0)^*E = 1^*E + 0^*E = E + Z.$$

PROPOSITION 7.7. $E(A,B)$ *is an abelian group under* (7.8).

PROOF. Plainly $E(A,B)$ is commutative and we identified the zero element in Proposition 7.6. It remains to prove associativity and the existence of inverses.

ASSOCIATIVITY. Let us write $\tilde{\Delta}: A \to A \oplus A \oplus A$, $\tilde{\nabla}: B \oplus B \oplus B \to B$ for the appropriate diagonal and codiagonal maps. Then $\tilde{\Delta} = (\Delta \oplus 1)\Delta$, $\tilde{\nabla} = \nabla(\nabla \oplus 1)$. Thus

$$\tilde{\nabla}_*\tilde{\Delta}^*(E_1 \oplus E_2 \oplus E_3) = \nabla_*(\nabla \oplus 1)_*\Delta^*(\Delta \oplus 1)^*(E_1 \oplus E_2 \oplus E_3)$$

$$= \nabla_*\Delta^*(\nabla \oplus 1)_*(\Delta \oplus 1)^*(E_1 \oplus E_2 \oplus E_3), \quad \text{by Theorem 6.7,}$$

$$= \nabla_*\Delta^*(\nabla \oplus 1)_*(\Delta^*(E_1 \oplus E_2) \oplus E_3), \quad \text{by (7.9)}$$

$$= \nabla_*\Delta^*(\nabla_*\Delta^*(E_1 \oplus E_2) \oplus E_3), \quad \text{by the dual of (7.9).}$$

$$= (E_1 + E_2) + E_3.$$

Similarly $\tilde{\nabla}_* \tilde{\Delta}^*(E_1 \oplus E_2 \oplus E_3) = E_1 + (E_2 + E_3)$.

EXISTENCE OF INVERSES.   We have

$$0 = 0^*E, \quad \text{by Proposition 7.6,}$$
$$= (1 + (-1))^*E$$
$$= E + (-1)^*E, \quad \text{by Proposition 7.5.}$$

Thus $(-1)^*E = -E$.

The reader will note how functoriality has been used essentially in the last argument. It would not be easy to find (or guess!) an explicit form for $-E$ and prove it to be the additive inverse of E.

The reader should also note that Theorem 7.2 and Proposition 7.3 together provide a proof of the fact that $\text{Ext}^1_R(A, B)$ is independent of the choice of projective resolution of A. Moreover, there is a dual development in which one obtains a natural equivalence $\bar{\theta}$ between $E(A, B)$ and $\overline{\text{Ext}}^1_R(A, B)$ and the analog of Proposition 7.3. Thereby one also obtains a proof of the natural equivalence of Ext^1_R and $\overline{\text{Ext}}^1_R$. As we have said, the arguments in this section may be generalized to Ext^n; the reader will find details in Section 9 of Chapter 4 of [4]; some indications are given in the Exercises.

EXERCISES

1.1. Show that $_R\mathfrak{Ch}$ is an abelian category.

1.2. Show that $\text{Hom}_R(A, C)$, $\text{Hom}_R(C, B)$, $A \otimes_R C$ are bifunctors. (See Examples 4, 5).

1.3. Let C be chain complex of right R-modules, let A be a left R-module, and let B be an abelian group. Establish a natural isomorphism

$$\text{Hom}(C \otimes_R A, B) \cong \text{Hom}_R(C, \text{Hom}(A, B)).$$

1.4. Similarly make sense of, and prove the existence of, natural isomorphisms

$$\text{Hom}(A \otimes_R C, B) \cong \text{Hom}_R(A, \text{Hom}(C, B)),$$
$$\text{Hom}(A \otimes_R B, C) \cong \text{Hom}_R(A, \text{Hom}(B, C)).$$

1.5. Let C be a chain complex of right R-modules and let D be a complex of left R-modules. Form the graded abelian group E such that

$$E_n = \bigoplus_{p+q=n} C_p \otimes_R D_q$$

and introduce the homomorphisms

$$\partial_n^{\otimes} : E_n \to E_{n-1},$$

given by

$$\partial_n^{\otimes}(x \otimes y) = \partial x \otimes y + (-1)^p x \otimes \partial y, \quad x \in C_p, \quad y \in D_q.$$

Show that $(\mathbf{E}, \partial^{\otimes})$ is a chain complex. Show that this construction generalizes Example 5, and that $(\mathbf{C}, \mathbf{D}) \mapsto \mathbf{E}$ yields a functor $\mathfrak{Ch}_R \times {}_R\mathfrak{Ch} \to {}_{\mathbf{Z}}\mathfrak{Ch}$. [We write $\mathbf{E} = \mathbf{C} \otimes_R \mathbf{D}$.]

1.6. Let $\mathbf{C} \in {}_R\mathfrak{Ch}_S, \mathbf{D} \in {}_S\mathfrak{Ch}_T, \mathbf{E} \in {}_T\mathfrak{Ch}_U$. Show that $\mathbf{C} \otimes_S \mathbf{D} \in {}_R\mathfrak{Ch}_T$ and establish a natural isomorphism of objects of ${}_R\mathfrak{Ch}_U$,

$$(\mathbf{C} \otimes_S \mathbf{D}) \otimes_T \mathbf{E} \cong \mathbf{C} \otimes_S (\mathbf{D} \otimes_T \mathbf{E}).$$

1.7. Establish a natural isomorphism

$$\mathbf{C} \otimes_R \mathbf{D} \cong \mathbf{D} \otimes_{R^0} \mathbf{C}.$$

1.8. Propose a definition of $\operatorname{Hom}_R(\mathbf{C}, \mathbf{D})$, generalizing Example 4 (both cases).

1.9. Let $\mathbf{C}, \mathbf{D}$ be chain complexes over $\mathbf{Z}$ defined as follows:

$$C_1 = \mathbf{Z} = (a); \quad C_0 = \mathbf{Z} = (b); \quad C_n = (0), \quad n \neq 0, 1; \quad \partial a = 2b,$$

$$D_1 = \mathbf{Z} = (a); \quad D_n = (0), \quad n \neq 1.$$

Let $\phi : \mathbf{C} \to \mathbf{D}$ be given by $\phi(a) = a$. Show that $\phi_* = 0$: $H(\mathbf{C}) \to H(\mathbf{D})$, but that $(\phi \otimes 1)_* : H(\mathbf{C} \otimes \mathbf{Z}_2) \to H(\mathbf{D} \otimes \mathbf{Z}_2)$ and $\operatorname{Hom}(\phi, 1) : H(\operatorname{Hom}(\mathbf{D}, \mathbf{Z}_2)) \to H(\operatorname{Hom}(\mathbf{C}, \mathbf{Z}_2))$ are nonzero.

2.1. Give an example of the following situation: $F : {}_R\mathfrak{M} \to {}_S\mathfrak{M}$ is an additive functor, $\mathbf{C}, \mathbf{D} \in {}_R\mathfrak{Ch}, \phi, \psi : \mathbf{C} \to \mathbf{D}, \phi_* = \psi_*, (F\phi)_* \neq (F\psi)_*$.

2.2. Give an example of the situation above with $\mathbf{C} = \mathbf{D}, \phi_* = 1$.

2.3. Show that, in the category ${}_{\mathbf{Z}}\mathfrak{Ch}$, the conclusion of Theorem 2.5 remains true without the restriction to *positive* chain complexes. [*Hint*: show that, given $H_* \in {}_{\mathbf{Z}}\mathfrak{M}^{\mathbf{Z}}$, there exists $\mathbf{P}$, pseudoprojective, with $H_*(\mathbf{P}) \cong H_*$. Then show that, given $\psi : H_*(\mathbf{P}) \to H_*(\mathbf{C})$, we can find $\phi : \mathbf{P} \to \mathbf{C}$ with $\phi_* = \psi$.]

2.4. Let $\phi : \mathbf{C} \to \mathbf{D}$ in ${}_R\mathfrak{Ch}$. Form the chain complex $\mathbf{E} = \mathbf{E}_\phi$ as follows:

$$E_n = C_{n-1} \oplus D_n, \partial(c, d) = -(\partial c, \phi c + \partial d).$$

Verify that $\mathbf{E}_\phi$ is a chain complex. Show that $H_*(\mathbf{E}) = 0$ if and only if $\phi_* : H_*(\mathbf{C}) \cong H_*(\mathbf{D})$.

2.5. Express the *functoriality* of E_ϕ in Exercise 2.4.

2.6. Give an example of a positive chain complex C such that $H_*(C) = 0$ but C is not contractible.

3.1. Establish (3.6) in detail.

3.2. Establish the naturality of the isomorphism of Theorem 3.3 (in A and B).

3.3. Compute $\operatorname{Ext}^n_{\mathbf{Z}}(A, B)$, where A, B are cyclic groups.

3.4. Show that $\operatorname{Ext}^n_R(A, B) = 0, n \geqslant 1$, if B is injective or if A is projective.

3.5. Prove Theorem 3.1*.

3.6. Show that, in the definition of $\operatorname{Ext}^n_R(A, B)$, we may replace A by an arbitrary positive chain complex without invalidating any steps in the procedure. Is there any corresponding generalization available with respect to B?

4.1. Use Theorem 4.2 to establish an exact sequence

$$\cdots \to H_n(\mathbf{C}) \xrightarrow{\phi_*} H_n(\mathbf{D}) \to H_n(\mathbf{E}) \to H_{n-1}(\mathbf{C}) \to \cdots$$

associated with a chain map $\phi : \mathbf{C} \to \mathbf{D}$, where $\mathbf{E} = \mathbf{E}_\phi$ is the chain complex described in Exercise 2.4. [*Hint*: exploit the short exact sequences $D_n \rightarrowtail E_n \twoheadrightarrow C_{n-1}$.] $\mathbf{E}_\phi$ is called the *mapping cone* of ϕ.

4.2. Prove that the exact sequence of Exercise 4.1 is natural.

4.3. (a) Given $A \xrightarrow{\phi} B \xrightarrow{\psi} C$ in $_R\mathfrak{M}$, establish an exact sequence

$$0 \to \ker\phi \to \ker\psi\phi \to \ker\psi \to \operatorname{coker}\phi \to \operatorname{coker}\psi\phi \to \operatorname{coker}\psi \to 0.$$

(b) Consider chain maps $\mathbf{C}' \xrightarrow{\phi} \mathbf{C} \xrightarrow{\psi} \mathbf{C}''$. Writing $H_n(\phi)$ for $H_n(\mathbf{E}_\phi)$, and so on, establish an exact sequence

$$\cdots \to H_n(\phi) \to H_n(\psi\phi) \to H_n(\psi) \to H_{n-1}(\phi) \to \cdots,$$

and prove its naturality.

4.4. Establish the six-term exact sequences of abelian groups, given short exact sequences $A' \rightarrowtail A \twoheadrightarrow A'', B' \rightarrowtail B \twoheadrightarrow B''$:

$$0 \to \operatorname{Hom}(A'', N) \to \operatorname{Hom}(A, N) \to \operatorname{Hom}(A', N)$$
$$\to \operatorname{Ext}^1(A'', N) \to \operatorname{Ext}^1(A, N) \to \operatorname{Ext}^1(A', N) \to 0.$$

$$0 \to \operatorname{Hom}(M, B') \to \operatorname{Hom}(M, B) \to \operatorname{Hom}(M, B'')$$
$$\to \operatorname{Ext}^1(M, B') \to \operatorname{Ext}^1(M, B) \to \operatorname{Ext}^1(M, B'') \to 0,$$

4.5. Use the second sequence of Exercise 4.4 to prove the following

(a) If A is torsion-free, $\operatorname{Ext}(A, \mathbf{Z})$ is divisible.

(b) If A is divisible, $\operatorname{Ext}(A, \mathbf{Z})$ is torsion-free.

4.6. Show that if A is not torsion-free, then $\operatorname{Ext}(A, \mathbf{Z}) \neq 0$.

4.7. (Harder) Show, conversely to Exercise 4.5, that

(a) if $\text{Ext}(A,\mathbf{Z})$ is divisible, then A is torsion-free;

(b) if $\text{Ext}(A,\mathbf{Z})$ is torsion-free and $\text{Hom}(A,\mathbf{Z})=0$, then A is divisible.

Show that we cannot discard the condition $\text{Hom}(A,\mathbf{Z})=0$ in (b).

4.8.   Verify that the isomorphism of Theorem 4.7 is actually an equivalence of bifunctors.

4.9.   Generalize Theorem 4.9 to homomorphisms $\phi:A \rightarrow A'$ inducing isomorphisms $\text{Ext}_R^n(A',B) \rightarrow \text{Ext}_R^n(A,B)$, for all B and fixed $n \geqslant 1$.

4.10. Show that a natural transformation of functors $\text{Ext}_R^1(A',-) \rightarrow \text{Ext}_R^1(A,-)$ is always induced by a homomorphism $\phi:A \rightarrow A'$. [*Hint*: apply the natural transformation to $[1_{Q'}]$, where $Q' \rightarrowtail P' \twoheadrightarrow A'$ is a presentation of A'.]

5.1.   Prove Theorem 5.4.

5.2.   Show that the isomorphism of Theorem 5.3 is natural in A and B.

5.3.   Show that $\text{Tor}_1^{\mathbf{Z}}(A,-)$ is a left exact functor $\mathfrak{Ab} \rightarrow \mathfrak{Ab}$.

5.4.   Evaluate $\text{Tor}_1^{\mathbf{Z}}(A,B)$ if A,B are cyclic groups.

5.5.   Show that $\text{Tor}_1^{\mathbf{Z}}(A,B)$ is always a torsion group. Show that A is a torsion group if and only if $\text{Tor}_1^{\mathbf{Z}}(A,\mathbf{Q}_1) \cong A$.

5.6.   Show that, if $\mathbf{C}$ is a chain complex of free abelian groups, and G is an abelian group, then there is a natural short exact sequence

$$0 \rightarrow H_n(\mathbf{C}) \otimes G \rightarrow H_n(\mathbf{C} \otimes G) \rightarrow \text{Tor}_1^{\mathbf{Z}}(H_{n-1}(\mathbf{C}),G) \rightarrow 0.$$

5.7.   Show that, under the same hypotheses as in Exercise 5.6, there is a natural short exact sequence

$$0 \rightarrow \text{Ext}_{\mathbf{Z}}^1(H_{n-1}(\mathbf{C}),G) \rightarrow H^n(\text{Hom}(\mathbf{C},G)) \rightarrow \text{Hom}(H_n(\mathbf{C}),G) \rightarrow 0.$$

5.8.   Show that the exact sequences of Exercises 5.6 and 5.7 split. [*Hint*: there are projections $C_n \rightarrow Z_n(\mathbf{C})$.] (It is interesting to note that the splitting is not natural; see [4] for examples.)

5.9.   Define, for a group G and G-module A, $H_n(G;A)=\text{Tor}_n^{\mathbf{Z}[G]}(\mathbf{Z},A)$, where $\mathbf{Z}$ has the structure of a trivial (right) G-module. We call $H_n(G;A)$ the nth *homology group of G with coefficients in the G-module A*. Evaluate $H_0(G;A), H_1(G;\mathbf{Z})$.

5.10. Show that $H_n(G;-)$ is an additive functor. (Harder) Describe the functorial nature of $H_n(-;A)$.

5.11. Propose a similar definition for $H^n(G;B)$. Obtain exact coefficient sequences for the homology and cohomology groups of the group G. Evaluate $H^0(G;B)$. Evaluate $H^1(G;B)$ when B is a trivial G-module.

6.1.   Provide an example to show that the conclusion of the converse part of Theorem 6.4 need not hold if ϕ_1 is not epic.

6.2. Prove Theorem 6.4*.

6.3. Show that if (6.1) is a pullback square, and if ψ_1 is the kernel of ξ, then ϕ_2 is the kernel of $\xi\psi_2$. Show, conversely, that if

$$
\begin{array}{ccc}
A_0 & & A_1 \\
\phi_2 \downarrow & & \downarrow \psi_1 \\
A_2 & \xrightarrow[\psi_2]{} & A
\end{array}
$$

is given and if there exists ξ such that ψ_1 is the kernel of ξ and ϕ_2 is the kernel of $\xi\psi_2$, then there exists a unique $\phi_1 : A_0 \to A_1$ completing a pullback square.

6.4. Consider the commutative diagram

$$
\begin{array}{ccccc}
A_0 & \overset{\epsilon_0}{\rightarrowtail} & M_0 & \overset{\mu_0}{\rightarrowtail} & B_0 \\
\alpha \downarrow & & \gamma \downarrow & & \downarrow \beta \\
A_1 & \underset{\epsilon_1}{\rightarrowtail} & M_1 & \underset{\mu_1}{\rightarrowtail} & B_1
\end{array}
$$

in $_R\mathfrak{M}$. Show that the composite square is a pullback if and only if each square is a pullback. What is the dual of this?

6.5. Consider the commutative diagram

$$
\begin{array}{ccccc}
A_0 & \overset{\phi_0}{\rightarrow} & B_0 & \overset{\psi_0}{\rightarrow} & C_0 \\
\alpha \downarrow & & \beta \downarrow & & \downarrow \gamma \\
A_1 & \underset{\phi_1}{\rightarrow} & B_1 & \underset{\psi_1}{\rightarrow} & C_1
\end{array}
$$

in $_R\mathfrak{M}$. Suppose that (i) the composite square is a pushout, (ii) the right-hand square is a pullback, and (iii) α is surjective. Show that the left-hand square is a pushout. What is the dual of this?

6.6. Verify that the "uniqueness of pushout up to canonical isomorphism" is precisely compatible with the equivalence relation on short exact sequences given by (6.3), so that β_* is indeed well-defined by $\beta_*(E) = {}_\beta E$.

6.7. Prove the dual of Theorem 6.8.

7.1. Let $A, B \in {}_R\mathfrak{M}$, and consider exact sequences $(n \geqslant 1)$

$$
E: \qquad 0 \to B \to E_n \to \cdots \to E_1 \to A \to 0
$$

in $_R\mathfrak{M}$. Declare $E \rightsquigarrow E'$ if there is a commutative diagram

$$
\begin{array}{ccccccccccc}
E: & 0 \to & B & \to & E_n & \to \cdots \to & E_1 & \to & A & \to 0 \\
& \downarrow & \| & & \downarrow & & \downarrow & & \| \\
E: & 0 \to & B & \to & E'_n & \to \cdots \to & E'_1 & \to & A & \to 0
\end{array}
$$

and let $\sim$ be the equivalence relation generated by $\rightsquigarrow$. Write **E** for the equivalence class of E, and let $\mathrm{Yext}^n_R(A,B)$ be the set of equivalence classes. Show that $\mathrm{Yext}^1_R(A,B) = E(A,B)$ and that $\mathrm{Yext}^n_R(-,-)$ is a bifunctor, contravariant in the first variable and covariant in the second. [*Hint*: define α^*, β_* as in the case $n=1$, and exploit Theorem 6.4.] ('Y' stands for Yoneda, who invented Yext.)

7.2. Let $\mathbf{P} \to A$ be a projective resolution of A so that $\mathrm{Ext}^n_R(A,B)$ $\cong \mathrm{Hom}_R(B_{n-1},B)/i^*\mathrm{Hom}_R(P_{n-1},B)$, naturally, where $n \geqslant 1$, and i is the embedding of $B_{n-1} = B_{n-1}(P)$ in P_{n-1}. Construct a commutative diagram

$$
\begin{array}{ccccccccccc}
P_{n+1} & \to & P_n & \to & P_{n-1} & \to \cdots \to & P_0 & \to & A & \to 0 \\
\downarrow & & \downarrow & & \downarrow & & \downarrow & & \| \\
\mathbf{E}: \quad 0 & \to & B & \to & E_n & \to \cdots \to & E_1 & \to & A & \to 0
\end{array}
$$

and hence a map $\phi : B_{n-1} \to B$. Show that the association $\mathbf{E} \mapsto \phi$ sets up a function $\theta : \mathrm{Yext}^n_R(A,B) \to \mathrm{Ext}^n_R(A,B)$. Conversely, given $\phi : B_{n-1} \to B$, construct the pushout

$$
\begin{array}{ccccc}
B_{n-1} & \overset{i}{\rightarrowtail} & P_{n-1} & \twoheadrightarrow & B_{n-2} \\
\phi \downarrow & \text{Pushout} & \downarrow & & \| \\
B & \rightarrowtail & E_n & \twoheadrightarrow & B_{n-2}
\end{array}
$$

and hence the exact sequence

$$
\mathbf{E}_\phi : \quad 0 \to B \to E_n \to P_{n-2} \to \cdots \to P_0 \to A \to 0.
$$

Show that $\phi \mapsto \mathbf{E}_\phi$ sets up a function $\bar{\theta} : \mathrm{Ext}^n_R(A,B) \to \mathrm{Yext}^n_R(A,B)$.

7.3. Show that $\theta, \bar{\theta}$ are mutual inverses.

7.4. Show that, by *splicing* exact sequences, we get a natural pairing

$$\text{Yext}_R^m (A, B) \times \text{Yext}_R^n (B, C) \to \text{Yext}_R^{m+n} (A, C).$$

To what does this correspond with regard to the Ext functor. Identify the case $n = 1$ with the connecting homomorphism.

7.5. Compute $E(A, B)$ where A, B are cyclic (abelian) groups. Describe explicitly the elements of $E(\mathbf{Z}_4, \mathbf{Z}_4), E(\mathbf{Z}_4, \mathbf{Z}_8), E(\mathbf{Z}_8, \mathbf{Z}_4), E(\mathbf{Z}_8, \mathbf{Z}_8)$.

7.6. Consider the diagram

$$D: \quad \begin{array}{c} \xrightarrow{\phi} \\ \downarrow \psi \end{array}$$

Show that

(a) If ϕ is injective and ψ is surjective, then D can be embedded in a a bicartesian square if and only if $\text{im} \phi \subseteq \ker \psi$.

(b) If ϕ is surjective and ψ is surjective, then D can be embedded in a bicartesian square if and only if the exact sequence [Exercise 4.3(a)]

$$\ker \phi \rightarrowtail \ker \psi \phi \twoheadrightarrow \ker \psi$$

splits.

(c) If ϕ is injective and ψ is injective, then D can be embedded in a bicartesian square if and only if the exact sequence

$$\text{coker} \phi \rightarrowtail \text{coker} \psi \phi \twoheadrightarrow \text{coker} \psi$$

splits.

(d) (Harder) If ϕ is surjective and ψ is injective, and if $R = \mathbf{Z}$, then D can always be embedded in a bicartesian square, but that this does not hold for general rings R.

LIST OF SYMBOLS

1. GROUPS AND MODULES

$\square\, G_i$	Smallest subgroup containing the subgroups G_i of a group G
ΣA_i	Smallest submodule containing the submodules A_i of a module A
$\bigoplus A_i$	Internal direct sum of submodules A_i of a module A
$\oplus A_i$	External direct sum of modules A_i
ΠA_i	Direct products of modules (or groups) A_i
$\mathrm{Aut}\, H$	Group of automorphisms of the group H
$H \times_\omega G$	Semidirect product of G and H with respect to an action $\omega : G \to \mathrm{Aut}(H)$ of G on H
A^ϕ	R-module resulting from turning the S-module A into an R-module via the (unitary) ring homomorphism $\phi : R \to S$
$\mathrm{Hom}_R(A,B)$	Group of R-homomorphisms from the R-module A to the R-module B
A^*	$\mathrm{Hom}_R(A,R)$
$A^\ddagger$	Left (right) R-module $\mathrm{Hom}(A,\mathbf{Q}_1)$, where A is a right (left) R-module
$\mathbf{Q}$	Group of rationals
$\mathbf{Q}_1$	Group of rationals *modulo* $\mathbf{Z}$
$\mathbf{Z}$	Group of integers
$\mathbf{Z}_n$	Group of integers *modulo* $n\mathbf{Z}$
$\mathbf{Z}^{(S)}$	Free group on the set S
$\mathbf{Z}^{[S]}$	Free abelian group on the set S
P_S	Group of bijective mappings from S to S

2. MORPHISMS

$\rightarrowtail$	Monic or monomorphism or injective function
$\twoheadrightarrow$	Epic or epimorphism or surjective function
$\rightarrowtail\!\!\!\twoheadrightarrow$	Monic and epic (hence isomorphism in $\mathfrak{G}$ or $_R\mathfrak{M}$)
$\cong$	Isomorphism
$\{f_i\}$	Unique morphism into a product determined by the

morphisms f_i

$\langle f_i \rangle$ Unique morphism coming out of a coproduct determined by f_i

$f]g$ f is the kernel of g

$f[g$ g is the cokernel of f

$f[]g$ $f]g$ and $f[g$

$\cdots\!\!\rightarrow$ Morphism whose existence is asserted or has been proved

$\overset{\psi}{\cdots\!\!\rightarrow}$ Morphism whose existence and uniqueness are asserted or have been proved

$\left. \begin{array}{c} a \overset{f}{\mapsto} b \\ f : a \mapsto b \end{array} \right\}$ Given function f maps element a to element b

$\phi \approx \psi$ Homotopy relation between chain maps ϕ and ψ

3. CATEGORICAL NOTATION

$F \overset{\eta}{-\!\!\dashv} G$ F is adjoint to G with adjugant η

$\{ \amalg A_i, q_i \}$ Coproduct of objects A_i with injections q_i

$\{ \Pi A_i, p_i \}$ Product of objects A_i with projections p_i

4. SPECIAL CATEGORIES

$\mathfrak{Ab}$ Category of abelian groups

$\mathfrak{C}^{op}, \mathfrak{C}^0$ Opposite category

$\mathfrak{C}_G$ Category with one object whose set of morphisms forms the group G under morphism composition

$\mathfrak{C}/Y$ Category whose objects are morphisms $X \rightarrow Y, X \in \mathfrak{C}$, for a fixed $Y \in \mathfrak{C}$

$_R\mathfrak{Ch}$ Category of chain complexes of left R-modules

$_R\mathfrak{Ch}_h$ Category of chain complexes of left R-modules and homotopy classes of chain maps; also called the *homotopy category* of $_R\mathfrak{Ch}$

$_R\mathfrak{Ch}_h^+$ Subcategory of $_R\mathfrak{Ch}_h$ consisting of positive chain complexes

$_R\mathfrak{P}_h^+$ Subcategory of $_R\mathfrak{Ch}_h^+$ consisting of pseudoprojective chain complexes

$\mathfrak{D}_K$ Category of K-vector spaces, or vector spaces over the field K

$\mathfrak{G}$ Category of groups

$_R\mathfrak{M}(\mathfrak{M}_R)$ Category of left (right) R-modules

$_R\mathfrak{M}^{\mathbf{Z}}$ Category of $\mathbf{Z}$-graded left R-modules

$_R\mathfrak{M}_S$ Category of (R, S)-bimodules

$\mathfrak{R}(\mathfrak{R}_1)$ Category of (unitary) rings

$\mathfrak{S}(\mathfrak{S}_0)$ Category of (pointed) sets

$\mathfrak{T}(\mathfrak{T}_0)$ Category of (pointed) topological spaces

BIBLIOGRAPHY

1. Bass, H., *Algebraic K-theory*, Benjamin, 1968.
2. Bourbaki, *Algèbre*, Hermann, Paris, 1948.
3. Cartan, H., and S. Eilenberg, *Homological Algebra*, Princeton University Press, 1956.
4. Hilton, P. J., and U. Stammbach, *A Course in Homological Algebra*, Springer-Verlag, 1970.
5. Lang, S., *Algebra*, Addison-Wesley, 1966.
6. MacLane, S., *Categories for the Working Mathematician*, Springer-Verlag, 1971.
7. MacLane, S., *Homology*, Academic Press, 1963.
8. Rotman, J., *The Theory of Groups*, Allyn and Bacon, 1965.

INDEX

Abelian group, 6, 38
  divisible, 50, 116
  finite, 86
  finitely generated, 42, 85
  free, 43
  injective, 48
  projective, 46
  torsion, 85
Abelianization, 69
Action, 30
Additive notation, 6
Adjoint functors, 98
Adjugant, 99
Adjunction, 99
  counit, rear of, 104
  unit, front of, 104
Alternating group, 36
Antisymmetric, 2
Ascending chain condition, 162
Associates, 131
  left, 131
  right, 131
Associativity, 3, 65
Automorphism, 30, 35
  inner, 35

Basis, 43, 145
  R-projective, 148
Bicartesian, 220
Bifunctor, 120
Bimodule, 137
Boolean algebra, 79
Boundary, 191
  operator, 191

Canonically equivalent, 81
Cartesian product, 1, 82
Category, 16, 64
  abelian, 109

additive, 107
of groups, 35
homotopy, 195
opposite, 74
quotient, 119
of sets, 66
small, 73
Center (of group), 12
Center (of ring), 156
Chain (ordered set), 2
Chain complex, 191
Chain homotopy, 194
Chain map, 191
Chain rule, 70
Characteristic polynomial, 180
Circle group, 50
Codomain, 65
Coefficients, 126
Coequalizer, 122
Cofree, 62, 153
Cohomology group (of group), 236
Coinitial (object), 67, 75
Cokernel, 20, 77, 133
Commutator, 33
Complement, 1
Complete set of representatives, 1
Complex, 191
  chain, 191
  cochain, 192
  positive chain, 193
  positive cochain, 200
  pseudoinjective cochain, 200
  pseudoprojective chain, 196
Complex numbers, 50
Component, 1, 84
Composition, 1, 65
Concentrated (in dimension 0), 193
Conjugate, 13
Contractible (chain complex), 195